D0670684

CHICAGO PUBLIC LIBRARY
R00681 53833
DISCARD

DATE

BUILT BY JAPAN

BUILT BY JAPAN

COMPETITIVE STRATEGIES OF THE JAPANESE CONSTRUCTION INDUSTRY

BY

FUMIO HASEGAWA

Massachusetts Institute of Technology

AND THE

SHIMIZU GROUP FS

Shimizu Corporation

WILEY

A WILEY-INTERSCIENCE PUBLICATION

JOHN WILEY & SONS

NEW YORK / CHICHESTER / BRISBANE / TORONTO / SINGAPORE

Library of Congress Cataloging-in-Publication Data:

Kensetsugyō no mirai senryaku. English.
Built by Japan: competitive strategies of the Japanese
construction industry / by Fumio Hasegawa and the Shimizu Group FS.
p. cm.
Translation of: Kensetsugyō no mirai senryaku. With new introd.
Includes bibliography.
ISBN 0-471-63254-6
1. Construction industry–Japan. 2. Strategic planning–Japan.
I. Hasegawa, Fumio. II. Title. III. Title: Built by Japan.
HD9715. J22K513 1988
624'. 068–dc19 88-20478
CIP

Printed in the United States of America

10 9 8 7 6 5 4 3 2

CONTENTS

PART TWO PASSPORT TO GROWTH—SIX STRATEGIES

PREFACE

All industries must experience a rise and a fall, good times and bad times, growth and decay. Quite a number of previously thriving industries have faded into minor operations hovering in a dim corner of the economy. Within such weak industries, however, some companies have found new markets and have evolved into worthwhile businesses. Obviously, with insight into the proper strategies, companies can avert the worst and again achieve growth even in the worst of times.

After a lengthy period of slow business growth, the Japanese construction market finally began to show brisk activity in the second half of 1986 as Japanese economic policy was shifted to derive a greater part of economic growth from domestic, rather than export, resources. Still, people in the construction industry are not overly optimistic, fearing that the current activity may not last for long.

This cautious attitude is a result of the Ice Age construction slump that continued for a decade from the mid-1970s to 1986 and is now in a temporary thaw period. That was a dramatic turnabout from the preceding construction boom which prevailed throughout the period of high economic growth from the late 1950s to the early 1970s, with the 1964 Tokyo Olympics as the super springboard. Because the Ice Age began so suddenly after the 1973 oil crisis, many Japanese construction companies failed to properly adapt to the slower business pace and have thus failed to chalk up increased profits.

Around 1985, we at Shimizu Corporation noted on one hand a strong, active Japanese economy which had successfully overcome two serious oil crises, and on the other hand an ailing construction industry which continued to slump, and we pondered why only the construction industry must play the role of odd man out in the industrial community. We were keenly aware that Japanese industry as a whole was being modified as part of a worldwide industrial reorganization movement, and we realized that the Japanese construction industry should be part of this megatrend if it was to enjoy the kind of prosperity achieved by other industries. Accordingly, we felt that Japanese contractors should try to expand their operations into foreign markets, while simultaneously an increasing number of foreign contractors were expected to establish their operations in the Japanese construction market.

At that time, we were approached by a Japanese publisher with a request to write a book on future business strategies needed by the construction industry, and since we had great interest in this subject matter, we accepted the offer and promptly formed a Group FS (Future Strategy) of young Shimizu managers who volunteered to be the co-authors of the planned book.

Although future projections are often seen through rose-colored glasses, in 1985 we were (and still are) deeply concerned with ways to escape from the Ice Age of the construction industry, and we have thus deliberately avoided painting a rosy picture of a twenty-first century full of marine and space structures, biotechnologies, and new construction materials bringing brand-new markets and opportunities. On the contrary, we consider that the remaining 15 years in the current century will be a crucial period for the Japanese construction industry.

Fortunately, the Japanese language version of this book, after publication in December 1985, appeared on the business book best sellers list for a number of months and has reached an impressive 18th reprinting to date. Readers were found not only in the construction industry, but also in other major industries, such as the steel and automobile industries. We interpreted this unexpectedly broad audience as the result of businessmen's appreciation of our determination to overcome general business slumps and achieve sustained growth, which is a common theme for all businesses.

Looking outward, we cannot ignore the growing foreign interest in the Japanese construction market, but people outside Japan have had only limited access to relevant information. What is the overall makeup of the Japanese construction industry? What problems do Japanese contractors face today? What courses do they intend to take in the future? Information on these and similar questions will certainly help overseas people to understand the construction business in Japan.

In this regard, convinced that our book contains sufficient information to fill up some of the information gaps in other countries, we decided to publish an English language version of our book, with a special introductory section presenting an overview of our industry for overseas readers, as well as some footnotes explaining important Japanese concepts. Most of the statistical data used in this English edition were updated from the original Japanese version. We hope that the English edition will serve foreign readers as a dependable guide to the Japanese construction industry and as a reference book for those seeking an understanding of present conditions in Japanese industries overall.

Additionally we have to explain the conversion rate from Japanese yen to U.S. dollar used throughout this book. As the currency is changing almost every day, we cannot fix it correctly. Only God knows it. In this book we considered it 150 yen to the U.S. dollar for reader's convenience.

Finally, we wish to extend our thanks to Mr. Teruzo Yoshino, president of Shimizu Corporation for his kind support of our project and to Mr. Kazuhiko Kondoh, general manager of Shimizu's corporate planning division, for his valuable advice. We also congratulate Mr. Yoshinori Oiwa of Alpha Business Co.,

Ltd. on his translation of this book, and his conquest of special Japanese construction terms. In addition, we are deeply grateful to Mr. Michael L. Joroff, who is the director of the Laboratory of Architecture and Planning at Massachusetts Institute of Technology, for his useful advice during the production of this book. We also wish to express our gratitude to Mr. Stephen A. Kliment, editor for John Wiley & Sons Publishers, Inc., and Everett W. Smethurst, former editor, for their editorial advice and the opportunity they offered to publish this book.

FUMIO HASEGAWA

Cambridge, Massachusetts
November, 1988

PROFILE OF AUTHOR AND CONTRIBUTORS

HASEGAWA, FUMIO (Group FS leader): Doctor in system engineering; analyst of social and technological trends; involved in the corporate planning division; currently visiting research scholar at the Laboratory of Architecture and Planning, Massachusetts Institute of Technology; directed the Book publication project and wrote Introduction and Epilog.

ASANO, SADAYASU: Social engineer engaged in the planning and promotion of technological development; wrote Chapter 7.

HIYAMA, TOSHIFUMI: Landscape designer involved in the planning and development of urban and regional development projects; wrote Chapter 6.

IWASAKI, HIROSHI: Economist assigned to the planning of marketing activities for the Japanese construction market; wrote Chapters 2 and 3.

IWASAKI, YO-ICHI: Architect active in the research and planning of advanced technologies; wrote Chapter 5.

KAWAZU, YUSUKE: Economist, serving as an assistant to vice-president; co-author of Chapter 1.

KIMURA, MAKOTO: Legal expert assigned to research and planning for new business development; currently attending American Graduate School of International Management; wrote Chapter 4.

KYOHYA NOBUO: Construction engineer serving as a construction site manager; wrote Chapter 2.

KOBAYASHI, TSUNEO: Social engineer involved in the planning and promotion of technological development; wrote Chapter 7.

MIYANOHARA, KUNIAKI: Economist active in company management planning; wrote Chapters 1, 4 and 8.

SUZUKI, KENJI: Economist engaged in market analysis for the design division and in the research and analysis of the construction industry; wrote Chapter 2 and Introduction.

YAMASHITA, MASAFUMI: Economist assigned to accounting; wrote Chapter 8.

YANO, TAKAYOSHI: International relations and political scientist serving as an assistant site manager of overseas construction project; currently attending Hong Kong Chinese University; wrote Chapter 8.

HAMAI, TATSUO (Group FS coordinator): Public relations and corporate communications manager.

MUTOH, KATZ-SUKE (Group FS coordinator): Legal expert serving as a marketing manager in the civil engineering division.

BUILT BY JAPAN

Sun Heights Arai-cho (Japan) Shimizu Corporation

INTRODUCTION

In recent years the United States and many other countries have shown an increased interest in the Japanese construction industry, and a large number of experts from construction companies, architectural firms, universities, and research institutes in the United States and elsewhere have visited Japan to carry out a firsthand, on-site investigation of the activities of Japanese construction companies. These visitors have expressed surprise and puzzlement at the different ways in which Japanese construction firms operate and manage their affairs. They have discovered, for example, that all large general contractors in Japan have their own high-level research laboratories staffed by a large number of top researchers. These labs were also impressively furnished with the latest scientific equipment, which is as advanced as that to be found in the research facilities of universities and government laboratories. One question asked by the visitors is, "Why do Japanese contractors need such elaborate laboratories of their own?" and there are many other questions to which the foreign visitors seek answers.

The answers can be obtained only by understanding the differences between the operations of U.S. and Japanese construction companies. In the United States, construction companies usually operate as "builders," responsible only for the construction of structures according to the architectural plans drawn up by independent architects. Japanese construction firms, on the other hand, operate in a number of different fields ranging from architectural design and construction work to city planning, engineering, and land development; they even engage in various non-construction activities. In addition, the division of Japanese construction firms into complex levels of contractors and subcontractors makes it difficult for Japanese and overseas people alike to fully understand the construction industry setup. This introduction, therefore, provides an overview of the Japanese construction industry in order to present readers with the basic knowledge needed to fully understand the following chapters.

CONSTRUCTION DEMAND AND INDUSTRIAL STRUCTURE

Construction investments in Japan totaled about 50 trillion yen ($330 billion)* in fiscal 1985,† an amount equivalent to 16% of the nation's Gross National Product of 320 trillion yen ($2,100 billion) for the same fiscal year. This ratio is

* The conversion rate used throughout this book is 150 yen to the U.S. dollar, or approximately the rate prevailing in 1987.

† The Japanese government fiscal year starts April 1 of the numerically same calendar year and ends on the last day of March in the following year. Accordingly, fiscal 1987 began on April 1, 1987 and ended on March 31, 1988. The majority of Japanese companies adopt this fiscal year for their own terms of business.

considerably higher than the corresponding ratio of less than 10% for the United States and most other industrialized countries. This phenomenon is attributable to the relatively underdeveloped state of infrastructures in Japan, such as parks, highways, bridges, airports, sewage systems, and housing conditions.

A breakdown of total construction orders shows that 60% of the monetary total comes from the private sector. The remaining 40% is placed by government offices. These figures have remained more or less constant for the past 10 years. Classified into types of work, the total amount of construction investments is private housing (29.8%), government civil engineering (26.4%), private non-housing (23.0%), and private civil engineering (13.3%). In all, the Japanese construction industry comprises 520,000 firms and employs 5.3 million people, providing jobs for a significant 9% of the total Japanese labor force.

Nevertheless, it would not be appropriate to regard all of the 520,000 firms as being on the same plane. The smallest contractors are one-man operations, and the largest contractors operate throughout Japan and in many parts of the world, constructing massive structures such as nuclear power plants and ultramodern highrise offices. The great majority of contractors are small, and their business status is far from secure. Nearly 90 percent of the contractors are capitalized at less than 10 million yen ($67,000), and nearly 50% are one-man operations with no other employees. Large- and medium-sized contractors are generally considered to be those capitalized at 50 million yen ($330,000) or more, but these larger contractors account for only 1.5% of the total number. Thus, to obtain an accurate understanding of the Japanese construction industry, it is important to grasp the extent of Japanese contractors' business operations, technical capabilities, and future strategies in terms of corporate size.

ZENECONS AS LEADERS

Officially, the construction business is divided into three groups according to the Japanese Contract Construction Business Law: general and specialized contractors, specialized subcontractors, and equipment installers. These three groups are further divided into a total of 28 subgroups. Dubbed *Zenecons* (Japanese acronym for *gene*ral *con*tractors), the general contractors are the dominant members of the Japanese construction industry. The Japanese general contractors, by definition, are capable of taking orders for and performing the entire range of construction and civil engineering work.* The general contractors can be further classified into

* The Japanese construction market is traditionally classified into the construction and civil engineering fields. University courses and academic societies are divided according to this division, and government construction departments are organized in line with these classifications. Most large contractors in Japan operate in both fields, although their business may be biased more toward one or the other.

"large," "semi-large," "medium-sized," "small," and "civil engineering oriented" groups, according to the amount of their sales. The five largest Zenecons are widely known as the *Big 5*, which comprises Shimizu Corporation, Kajima Corporation, Taisei Corporation, Takenaka Corporation, and Ohbayashi Corporation. Recently, Kumagai Gumi Company has often been included in the top group, making it the Big 6.

Although capable of providing clients with the entire range of necessary services, general contractors usually subcontract work to specialized subcontractors and equipment installers. Construction work is carried out by a wide variety of subcontractors, and in many cases these subcontractors employ their own subcontractors. Subcontracting is less frequent in civil engineering work, partly because it requires fewer work specialists and partly because the use of machinery is more advanced in civil engineering. In the civil engineering field, even small firms often have a command of advanced technologies and are capable of acting as contractors.

The Japanese subcontracting system is based not only on tradition but also on the characteristic operations of general contractors, which resemble the operations of general trading companies. Like the giant trading houses, the Japanese large general contractors have always acted as project organizers, financiers, and managers. Also, because of the practice of employing workers on a lifetime basis, the general contractors consider it more advantageous to hire subcontractors on a project-by-project basis than to hire work specialists as permanent employees. By hiring subcontractors, the general contractors are able to adjust their work force in keeping with market ups and downs and to minimize the problems of transferring workers from one region to another whenever a new project starts. This practice saves manpower costs and ensures employee satisfaction, and accordingly, general contractors try to foster a family of subcontractors in each region.

THE HISTORICAL BACKGROUND OF CONTRACTORS

Construction activities in Japan are as old as its history. The world's oldest remaining wooden building, the Horyuji Temple in Nara, was built in the seventh century, and Japan can boast of many sophisticated temples, shrines, and emperors' massive mausoleums built in ancient times. But these structures were built under the direct leadership of emperors and shoguns, and "contractors" were still nonexistent in those times. The emergence of contractors began in the Edo Period (1603–1868), which featured the relocation of the capital from Kyoto to Tokyo and a rapid development of cities. Edo, the old name for Tokyo, became the world's largest city by the seventeenth century, probably the first city of one million people on earth. As a result, some of the carpenters who built mansions for samurais and stores for merchants gradually became contractors, receiving the whole order in exchange for a guarantee to complete the work. Those carpenter bosses of the

Edo Period are regarded as the forerunners of the present-day contractors in the Japanese construction industry.

At the beginning of the Meiji Period in 1868, Western architecture was introduced into Japan, and some of the carpenter bosses from the Edo Period were eager to master Western architectural and construction techniques. Kisuke Shimizu II (whose father was the founder of Shimizu Corporation) was among the internationally minded, and in 1867 built the Western-style Tsukiji Hotel in Tokyo. Iwakichi Kajima, the founder of Kajima Corporation, was the contractor who built the British First Mansion and the American First Mansion in Yokohama's foreign settlement site near Tokyo. Kisuke Shimizu and his compatriots acquired a basic knowledge of Western architecture through business contacts with the foreign architects who oversaw the construction of the buildings being erected in Japan by their governments and corporations. Thus, the design and construction of Western-style buildings by contractors began in the Meiji Period, roughly 100 years ago.

The growth of these early contractors was attributable largely to increases in government construction orders. The Meiji government pursued an aggressive policy of building administrative offices, railroads, stations, and other infrastructures as part of its effort to modernize Japan. At first, the government undertook the construction projects as a sort of self-appointed contractor, hiring foreign architects and managers, but as the volume of necessary construction work expanded, government contracts were given to private contractors. It was in this early Meiji Period that many of today's leading contractors came into existence, thanks to the multitude of government orders.

Following the patterns laid down during the Meiji Period, the development of the modern Japanese construction industry was based on Western technologies and the increased government and private orders resulting from the growth of Japan as a modern state. In line with this process of expansion, leading contractors were reorganized into modern corporations. Takenaka became an incorporated entity in 1909, Shimizu in 1912, and Kajima in 1930.

CHARACTERISTICS OF THE JAPANESE CONSTRUCTION INDUSTRY

Large Domestic Market

The Japanese construction industry differs from those in Western countries in a number of ways, both culturally and historically. First, the Japanese industry has a large domestic market, upon which it depends heavily. As mentioned before, construction investments in Japan are large in relation to the GNP partly because its infrastructure, including housing conditions, is comparatively underdeveloped. Today, demand is growing for a substantial elevation of the nation's infrastructure level, as a principal means to stimulate Japanese domestic demand and thus alleviate trade collisions with other countries.

According to an international study compiled by the Japanese Construction Ministry, sewage systems serve only 34% of the population in Japan, compared to 72% in the United States, 97% in Britain, 91% in West Germany, and 65% in France. Per capita area of city parks amount to only 2.1 square meters in Tokyo, compared to 19.2 square meters in New York City, 30.4 square meters in London, 37.4 square meters in Bonn, and 12.2 square meters in Paris. In addition, Japan lags behind in the development of highway and expressway systems, in the quality of housing, and in the structural appearances of cities. As a result, Japanese construction companies have always had an expanding domestic market to lean on, whereas contractors in Western countries have been compelled to expand into the international markets to cope with their sluggish domestic markets.

Construction Licensing and Bidding Systems

Another characteristic of the Japanese construction industry is that the Contract Construction Business Law requires prospective contractors to acquire an official license in order to start a construction business. The types of licenses differ according to the scale of the intended construction business, such as the locations of main and branch offices and the capability to hire a certain number of subcontractors. The licenses for large general contractors are issued by the Construction Minister. It is, of course, possible for foreign construction companies to obtain these business permits, provided that certain requirements are met; for example, the ownership of a minimum amount of property and the employment of a minimum number of qualified construction engineers in Japan.

Another important feature of the Japanese construction industry is that orders by the central and local governments for public construction work are placed, in principle, through public bidding. In evaluating the bids, the central and local governments rank construction firms according to past order-taking performance, sales, financial status, and technological capabilities. As a rule, Japanese public offices place orders following a designated competitive tendering system whereby selected ranks of construction companies are invited to take part in bidding according to the nature of the public works project. (Bidding in the United States is based on an open bid system.) If foreign contractors want to win orders for large projects from Japanese administrative offices, they must first prove their work capabilities by performing smaller projects. Companies found to have engaged in bribery or to have been involved in major accidents in public works projects are usually barred from public bidding for a fixed period of time.

In private orders, the client either appoints a specific contractor or invites a number of specific contractors to estimate prices for the proposed project. The percentage of the former type of special appointment contract, which does not require bidding, has decreased in recent years, and that of the latter type, where contractors are pitted against each other, has increased, often putting pressure on contractors to cut prices to such an extent that only a bare minimum profit can be expected.

Absence of Strong Unions

In contrast to the situation in the United States, where contractors often acquire construction labor by signing a contract with labor unions, Japanese construction companies do not depend on unions for the supply of labor. That is because, rather than maintain a work force of their own, Japanese contractors hire subconstractors on a regular basis, and these subcontractors are not unionized. Trade unions, such as the unions of carpenters, plasterers, and masons, do not exist in Japan.

Distinction between Construction and Civil Engineering

It is customary in Japan to divide construction activities into *construction* and *civil engineering*. The former includes the construction of houses, office buildings, hospitals, factories, and other buildings, and the latter includes site renovation work and the construction of dams, roads, highways, tunnels, bridges, and other large structures. General contractors operate in both fields, and they regard urban redevelopment and regional development projects as a most promising market area because these projects require both construction and civil engineering operations.

The U.S. industry does not make a distinction between construction and civil engineering. For example, in its regular listing of the top 400 construction firms, *Engineering New Record (ENR)* divides the construction business into nine categories according to the type of structure and operation. These nine categories are building (construction of houses, offices, stores), manufacturing (manufacturing facilities), power (power facilities), airport, highway and bridge, process (plant engineering), marine (marine structures), design, and construction management. However, since projects are becoming increasingly complex and extensive, in the future the Japanese two-field classification may be replaced by a more detailed classification, such as the one adopted by *ENR*.

Integration of Design and Building

Automakers, consumer electronics product makers, and most other manufacturers perform the whole range of product operation, from merchandise planning, design, and production to marketing. Yet in the construction business in the West, design is usually done by an independent architect or a consultant office, and the construction companies specialize in the building work. The underlying principle of this practice is that the quality of the constructed product can be better ensured by letting the designer and constructor check each other's work. Buildings being long-term-use products, the true qualities cannot be accurately assessed by the initial finish. It seems that the designer-constructor separation has developed from a desire to ensure high-quality buildings able to stand the test of time by applying a strict quality check throughtout the construction work process.

Nevertheless, construction companies in Japan usually carry out both design and construction work, and this is particularly true for private projects. This dual performance is probably attributable to a sense of "confidence" between the client and the contractor, consolidated over many years. As suggested by the unity of design and production taken for granted in manufacturing industries, it may be more natural from the standpoint of corporate management if construction companies perform both design and construction work. This will, once and for all, make the construction company itself responsible for the quality of its product and enable the company to better perform the budgetary planning, production control, and maintenance service that will most benefit the clients. Cultural and traditional factors are particularly strong in this specific issue, and thus the independence of design from construction, regarded as a matter of course in the West, is not so in Japan.

To integrate design and construction work, Shimizu, Kajima, Takenaka, and other large Japanese general contractors utilize design departments staffed by as many as 1000 architects and structural, mechanical, electrical, and other engineers, and these design departments compare favorably with leading specialized architectural firms in scale and scope of business activities. In addition, the large general contractors employ many design engineers in their civil engineering divisions, although not as many as they need for their construction divisions. This is because the majority of civil engineering orders are placed by government offices, which use their own design sections or hire specialized consultant firms to plan civil engineering projects.

The design departments are always involved in the projects contracted by the marketing departments, and therefore, must handle many diverse construction orders. The design departments are also actively engaged in the development of new design technologies, such as CAD (computer aided design) systems. The design department makes a vital contribution to the company management by adopting new technologies for practical use (along with the research and development department) and by reducing construction costs and upgrading product quality hand in hand with the construction work department.

Relations between Zenecons and Designers and Consultants

General contractors alone do not control the architectural design market, but operate side by side with specialized architectural and consulting firms and individuals (see Tables 1 and 2). These firms and individuals are well established, and

Table 1 Ranking of Japanese Architectural Firms by Number of Employees

Rank	Firm	Number of Employees[a]
1	Nikken Sekkei	1,338
2	Mitsubishi Estate Co. (design dept.)	494
3	Kume Architects & Engineers Co.	490
4	Nihon Architects, Engineers & Consultants	445
5	Yamashita Architects & Engineers	409
6	Yasui Architects Co.	346
7	Kozo Keikaku Engineering	340
8	Azusa Sekkei Co.	330
9	Rui Sekkeishitsu Co.	322
10	Ishimoto Architectural & Engineering Firm	272

[a] Data as of April 1, 1987.

Source: Nihon Keizai Shinbunsha, "Construction and Architectural Design Corporate Management Survey," *Nikkei Architecture,* Nihon Keiza: Shinbunsha, July 13, 1987, 70.

some of the architects have won worldwide acclaim. Observers from outside Japan may find it difficult to understand how these specialized firms and professionals can exist when general contractors command so many architects and architectural engineers of their own, but the explanation is simple: Although construction projects in Japan usually involve contractors' in-house design staff (i.e., design-

Table 2 Ranking of Japanese Civil Engineering Consultant Firms by Number of Employees

Rank	Firm	Number of Employees[a]
1	Nippon Koei Co.	1,586
2	Kokusai Kogyo Co.	946
3	Pacific Consultants K.K.	709
4	Ohba Co.	631
5	Hokkaido Engineering Consultants Co.	566
6	Tokyo Electric Power Service Co.	557
7	CTI Engineering Co.	556
8	Yachiyo Engineering Co.	543
9	Nihon Suido Consultants Co.	515
10	The New Japan Engineering Consultants	463

[a] Data as of September, 1987.

Source: Construction Directory 1988 (Kensetsu Meikan 1988), (Tokyo: Nikkan Kensetsu Tshushinsha, 1988).

build, turn-key, etc.), it is also a common practice to hire independent architectural firms and individuals as in the West.

General contractors are not necessarily the rivals of independent architects and consultants; there is as much cooperation as competition between these parties. For example, general contractors may be invited to take part in architectural firms' design work, or conversely, general contractors may invite the independents to join in design assignments. Recently, a number of famous foreign architects were hired by general contractors to produce design contest plans for large construction projects, working with the design departments. In the future, such cooperation between general contractors and independents is expected to increase in urban redevelopment and other large projects, not only because of the need to produce better designs but also because of the growing need for sophisticated technologies for installations and structures.

The picture is slightly different in the civil engineering field, in that the main clients—the government offices—have their own knowledgeable engineers to plan, design, and oversee projects, and civil engineering consultants are usually hired only to support the government engineers. The involvement and responsibility of these private consultants and also construction companies in public projects are nevertheless gradually expanding, and especially when a new, difficult technology is required, large construction companies are invited to assist in the planning, design, and testing of such a project.

On the overseas front, the cooperation between Japanese construction companies and foreign architectural and consulting firms is expanding mainly in two ways. First, the Japanese are collaborating with Western architectural and consulting firms in large projects for developing countries. Second, the Japanese are cooperating with local architectural and consulting firms of host countries; this is rapidly becoming a popular practice for projects in the United States, Southeast Asia, and Europe. The main aims are to implement projects with a minimum number of Japanese personnel, whose relative salaries are increasing due to the recent yen upvaluation, and to employ a maximum proportion of local personnel, who are naturally more knowledgeable of local legal requirements, customs, and ways of getting things done in the particular country. For these reasons, Japanese cooperation with local overseas architects, consultants, and engineers is most likely to expand even further in the future.

Active Involvement in Research and Development

Japanese construction firms, especially the general contractors, are constantly looking for new markets and new research and development (R&D) themes. They are particularly keen about advanced technologies, such as the air purification expertise applied to high-tech clean rooms for the production of semiconductor chips. Since orders for high-tech clean rooms cannot be captured without mastering the related advanced technologies, leading general contractors are forging ahead with their research and development efforts. Similarly, they are engaged in research and development activities to acquire, for example, space technologies, in hopes of winning orders for space structures in the future. In 1987, Shimizu Corporation became the first Japanese general contractor to set up a section specializing in space development. All large general contractors in Japan have high-level laboratories and research staffs of their own to push ambitious R&D programs. This active involvement in research and development may be a sign of the staunch entrepreneurship of Japanese contractors and their determination to cope with changes in market conditions.

PART ONE
ESCAPE FROM THE ICE AGE

Tuen Mun Highway and Interchange (Hong Kong) Aoki Corporation

CHAPTER 1

PLANNING FOR THE FUTURE

THE NEED TO RESTRUCTURE

The exchange rate of the Japanese yen against the U.S. dollar has soared to dizzying heights since the fall of 1986; less than two years previous to this writing, the rate stood at 240 yen to the dollar but now is only 125 yen, and higher rates loom closer every day. At 240 yen to the dollar, a Japanese exporter who had earned $10,000 received 2.4 million yen in exchange. Today, such an exchange will net only 1.25 million, or 48% less for the same amount of dollars.

This, plus foreign restrictive moves against Japanese exports, has dealt a hard blow to Japanese industries, especially the export-oriented automobile and electronic machinery industries. After a decade of prosperity as the elite among the nation's industries, automakers and electronics companies are now suffering from fast-dwindling profits. The steel and shipbuilding industries, which have suffered hardships brought about by the newly industrializing countries for a number of years, are now under additional stress due to the high yen.

The Japanese economy as a whole is under pressure and must restructure itself to become less dependent on export business and to build up its home market demand. Voices are raised both inside and outside Japan about a stimulation and expansion of domestic demand, and the construction industry is likely to benefit from this new development. But the basic outlook of the construction industry is and will be pessimistic, as a long-range look into the past and present clearly shows.

The Ex-Prince

Not long ago, the word *recession*, or even *slack*, did not exist in the vocabulary of Japanese contractors. When the Japanese economy expanded at a miraculous speed for about 15 years from the late 1950s to the first oil crisis in 1973, construction and civil engineering projects abounded throughout Japan. Producers of oil products, petrochemical, steel, and other intermediate industrial materials built massive plants on coastal sites; many big companies demanded larger high-rise office buildings to house an increasing staff; and service companies wanted bigger, shinier hotels, department stores, and other structures to attract the newly affluent consumer. In the public sector, the demand for dams, highways, harbors, and express railroads was strong everywhere in Japan. In those years, the total volume of construction works expanded more than 20% per year, exceeding the economic growth rate.

Almost all contractors, large or small, made handsome gains from the overflow of demand for construction services. General contractors, who are the largest contractors, became particularly prosperous and began appearing in the ranks of top corporate income earners of all industries. During the eight-year period from 1968 to 1975, for example, a cumulative total of 33 contractors made it into the top 50 annual income rankings, and the construction industry rose to be the fourth most prosperous after the finance, electronic machinery, and transportation machinery industries (see Table 1.1).

A dramatic change was brought about by the 1973 oil crisis and the ensuing economic slump. The growth in the volume of construction projects came to a screeching halt, causing hot competition among contractors. In the following years, many contractors were forced out of business, and a great majority of the survivors suffered from diminishing profits. This stumble by the construction industry was most visible in the aforementioned top 50 income ranking. Compared to the cumulative total of 33 contractors ranked among the top 50 income earners for the 1968–1975 period, there were only four contractors in the eight years from 1976 to 1983. Twenty-nine companies had disappeared from the ranking between the two eight-year periods, and no other industry lost so many top-ranked companies as the construction industry.

Table 1.1 Top 50 Companies in Corporate Annual Income

Industry	Number of Firms (1968–1975)	Number of Firms (1976–1983)	Change in Number
Finance	122	125	+ 3
Electric machinery	50	68	+18
Transportation machinery	37	30	− 7
Construction	**33**	**4**	**−29**
Utilities	33	48	+15
Steel	30	19	−11
Trading	19	23	+ 4
Food	13	14	+ 1
Chemical	11	14	+ 3
Mining	11	16	+ 5
Textile	11	0	−11
Rubber	8	4	− 4
Services	7	5	− 2
Industrial machinery	7	11	+ 4
Cement	6	1	− 5
Oil refining	1	16	+15
Others	1	2	+ 1
TOTAL	400	400	

Source: Produced from *Economic Statistics Annals (Keizai Tokei Nenkan)*, 1968–1983 editions (Tokyo: Toyo Keizai Shinposha).

Slump in Private and Public Sectors

The construction industry was probably hit the hardest by the oil crisis, and contractors were not indulging in hyperbole when they dubbed the dark mid-1970s the beginning of an Ice Age. A slump in the demand for construction work persisted for a long time because companies in most other industries placed a curb on plant and equipment investments, consumers postponed plans to buy new homes, and government offices scaled down public works projects. The construction industry thus lost all of the growth ingredients it had so long taken for granted during the period of high economic growth.

On the private plant and equipment front, smaller contractors complained that big general contractors were bidding for the minor orders they had hitherto ignored. This happened because the soaring price of oil and other fuels had rapidly decreased the need for energy-guzzling giant plants, such as blast furnace steel works and petrochemical complexes. The Japanese industrial structure as a whole turned to smaller, more energy-efficient factories and products. Corporate executives subscribed to defensive policies, and rather than build new factories and offices, managers purchased higher-grade automated machinery to reduce the labor force and installed more R&D equipment to prepare for the coming high-tech era.

As a result, Japan's industrial structure as a whole was reshaped in such a way that more industrial and electrical machinery was required but fewer construction orders were placed. The new industrial structure increasingly favored smaller construction projects. Also, the differential in the technical capabilities of large, medium, and small contractors lessened, because of the spread of advanced technologies, except in the fields of nuclear plants, super–high-rise buildings and a few other types of structures where large contractors predominate. Thus, it was no longer possible for the big, medium, and small contractors to coexist and share the market.

A slump also occurred in the house-building market, where small and medium contractors are most active. During the period of high economic growth, the number of home construction starts showed a large increase because of an acute home shortage, but in 1973, for the first time the number of housing units exceeded the number of households in all regions including big city areas in Japan. The number of homes constructed reached a peak in fiscal 1972 with 1.85 million housing starts, but then the increase in the number of households slackened while house and land prices began to rise much faster than the income increases during the inflation period following the 1973 oil crisis. The market for single-unit homes on a plot of land, where many small local home builders operate, became particularly sluggish.

To fight the recession after the oil crisis, public works projects were maintained at a high level, and this shored up the construction industry to a considerable extent.

By 1977 or 1978, public works were still effective in stimulating the economy, but were no longer strong enough to offset the construction slump resulting from reduced private plant and equipment investments. Crippled by persistent shortages in the government budget, public works completely lost their impact as a prime mover in the 1980s. The squeeze on public works was felt most by local contractors operating mainly in the civil engineering field. Public projects became so few that government administrators had to impose a "joint venture" formula, whereby large and smaller contractors were made to work together and share public works projects, mainly to help small local construction firms. Yet, despite these efforts, the financial status of the local contractors remained shaky.

Backwash of Overgrowth

Of course, contractors did not just stand with arms akimbo and watch their businesses dwindle and profits shrink. General contractors tried to expand by branching into the development business, engineering operations, and the overseas market, among other endeavors, and they launched total quality control (TQC)* programs and manpower reduction measures. As a result of these efforts, construction orders from abroad for the first time topped the one trillion yen mark ($6.7 billion) in fiscal 1983; this was a 5.9-fold expansion in 10 years.

Still, the domestic market did not improve. Rather, it worsened. Although the volume of construction work failed to increase for the 10 years following the 1973 crisis, the number of contractors increased by a whopping 70%. During the same 10-year period, the number of workers in the construction industry grew by 11.6% to 5.41 million, absorbing farm workers and the factory workers who were jobless because manufacturers had executed belt-tightening programs. One reason for the increase in the number of construction companies is that construction is a relatively easy business to break into for newcomers. A major reason for the influx of workers from other industries is that the construction industry lacked the keen sense of crisis harbored by other industries, and therefore failed to make wholehearted efforts to improve productivity.

Today over half a million construction firms are competing for slices of a relatively smaller pie. Managers of the 43 leading general contractors who form the Japan Federation of Construction Contractors lament that their companies can handle 50% more work with their present capacities, pointing to an acute shortage

* Total quality control is a corporate management technique based on the concept of quality control advocated by Dr. W. Edwards Deming of the United States. It features a means of solving quality problems by scientific techniques and companywide participation by all employees from the president down to each worker. Introduced to Japan in the 1950s, it was first practiced by manufacturing industries, and then, in the mid-1970s, was taken up by the construction and service industries.

of work and a large idle capacity. An average of 1000 to 2000 construction firms went out of business every year during the high economic growth period, but the number climbed to about 5000 per year in 1983 and reached 6000 in 1984. Industry analysts warn that many smaller contractors will soon be in the red if the market slump continues.

Today, however, there is a ray of hope. Because of the growing political need for a stimulation of domestic demand, to reduce both the nation's large trade surplus and trade collisions with other countries, renewed attention is being paid to the importance of construction activities. This has at last boosted the number of planned construction projects. Thanks to lowered interest rates and expanded housing loans, the number of housing starts increased to an estimated 1.4 million units in fiscal 1986. Public works projects are also increasing, since the government has decided to allocate more money to this section of the budget. Private construction work, too, is on the move, led by an increased demand for office buildings in big cities. Nevertheless, the Japanese construction industry remains basically oversized for the modest growth potential of the market, and the long-range viewpoint is not optimistic.

GAZE INTO THE FUTURE

What will be the future condition of the construction industry? Will the crisis linger? What will the future environment be like for contractors?

Little Prospect of Market Expansion

In 1984 the Construction Ministry inaugurated a Construction Industry Vision Study Group consisting of construction experts in the industrial, governmental, and academic fields. The task of this group was to debate and propose the future orientation of the construction industry. In February 1986, the group produced a final report entitled "A Vision of the Construction Industry in the 21st Century—The Challenge of Building a Vigorous Industry." This report attempted a long-term projection of the construction market, assuming three hypothetical cases of real growth in public-sector investments between 1983 and 2000: 0%, 3%, and 5% per year. This assumption was made in view of the sharp fluctuations in public investment amounts contingent on policy changes. The group also assumed that the Japanese GNP will increase by an average 4% per year in real terms during the 1983–2000 period (see Table 1.2).

According to the group's projection, Japan's total construction investments from 1983 to 2000 will expand at an average rate of 1.9% to 3.7% per year. These growth rates will be clearly higher than the zero construction growth actually

Table 1.2 Long-Term Projection of Demand for Construction Works (Unit: billion yen)[a]

Demand	Total	Growth[c] 1983–1990	Construction	Share	Civil engineering	Share	Ratio to GNP
1983							
Case[b] I, II, III	38,668		25,001	64.7%	13,668	35.3%	18.3%
1990							
Case I	46,681	2.7%	31,030	66.5%	15,651	33.5%	16.6%
II	48,377	3.3%	31,645	65.4%	16,732	34.6%	17.2%
III	44,413	2.0%	30.243	68.1%	14,170	31.9%	15.9%
2000							
Case I	62,806	2.9%	42,768	68.1%	20,037	31.9%	14.7%
II	72,066	3.7%	46,953	65.2%	25,113	34.8%	16.1%
III	52,835	1.9%	38,261	72.4%	14,575	27.6%	13.2%

[a] All monetary figures are based on 1975 prices.

[b] Case I assumes an average annual growth of 3% in public works investments between 1983 and 2000; the corresponding rates of 5% and 0% are assumed in Case II and in Case III, respectively. Japan's average GNP growth between 1983 and 2000 is assumed to be 4.2% per year for Case I, 4.5% per year for Case II, and 3.9% per year for Case III.

[c] Growth rate per year.

Source: A Vision of the Construction Industry in the 21st Century (Niju'uisseiki eno Kensetsu Sangyo Bijyon), (Tokyo: Ministry of Construction and the Construction Industry Vision Study Group, 1986).

registered during the 10-year period from the mid-1970s to the mid-1980s. Yet the group predicted that the construction growth will be lower than the predicted GNP growth by about 0.8 to 2 points, so that the ratio of construction investment to GNP will drop to between 13% and 16% by the year 2000, from the 18.3% actually recorded for fiscal 1983. The corresponding ratios for other industrialized countries are currently around 10%: for example, United States 7.9%, Britain 8.1%, and West Germany 12.5% (see Table 1.3). The reasons for Japan's higher construction investment-to-GNP ratio compared to those of Western countries are a still lower level of social capital stock, such as highway and sewage systems, and a higher level of economic growth. The weight of construction activities in the Japanese economy, however, will be certain to follow the patterns of Western countries, thus giving more reasons for Japanese contractors to take a serious view of the future.

Table 1.3 Construction Industries of Principal Countries

	Japan	United States	Britain	W. Germany
GNP (1983)[a]	¥274,639,000	$3,311,000	£3,011	DM1,671,000
Construction demand (A)[a]	¥47,980,000	$263,000	£243[b]	DM209,000
Ratio of (A) to GNP	17.5%	7.9%	8.1%	12.5%
Number of construction establishments	550,469 (1981)	1,200,407 (1977)	160,596 (1983)	72,606 (1983)
Percentage with fewer than 10 workers[c]	76.9%	92.6%	89.3%	54.9%
Change in number	Increasing	Increasing	Increasing	Leveling off
Number of construction workers	5,410,000 (1982)	5,290,000 (1977)	1,650,000 (1983)	1,450,000 (1983)
Change in number	Leveling off	Increasing	Decreasing	Decreasing

[a] Unit : millions

[b] Including demand for repair work.

[c] Percentage with fewer than 7 workers for Britain.

Source: Construction Industries of Major Western Countries (Shuyo'okoku no Kensetsugyo'o no Genjo'o to Do'oko'o), Research Institute of Construction and Economy.

The Need for Sophisticated Construction

High-tech clean rooms, intelligent buildings, and century-lasting houses are some of the new types of structures considered to have a large market potential, and leading construction firms are eager to master these new technologies. High-tech *clean rooms* are furnished with air-harmonizing equipment to keep dust, particle-laden air, temperature, humidity, air currents, and pressure differentials to a certain level. The first clean rooms in Japan were built around 1960, to house precision machine assembly lines, and recently a large number of more sophisticated clean rooms have been constructed for companies in the electronics and biotechnology fields. The clean rooms for semiconductor plants were at a Class 100 precision level in 1960, meaning a maximum 100 dust particles with a diameter of 0.5 micron or more per cubic foot, but today, to the production of 256-Kilobit to 1-Megabit super LSIs, a Class 10 precision level (a maximum of 10 particles per cubic foot) is required, with a maximum particle diameter of only 0.1 micron.

There is now a growing need for clean rooms that can be easily reassembled, to suit the rapid advances in the technical requirements of customers. In addition, customers want clean rooms featuring energy savings and low construction costs. The use of clean rooms is spreading from semiconductor and precision machinery production to hospitals and biotechnological labs for pharmaceutical, food, and agricultural products. If contractors neglect the diversified and sophisticated requirements of users for clean rooms, they will lose out to the competition.

Intelligent buildings were created in the United States to give tenants ready access to a high-level information and communications system and thus upgrade office productivity. In Japan also, office users look for a quick communications network system linking them directly with customers and branch offices to improve business efficiency. They also expect intelligent buildings to improve the safety and comfort of office life. Office users now consider it important to have a high-tech office in order to be successful in this new age of information. Large and medium contractors in Japan have already begun their attempts to master the intelligent building technologies in a bid to be first in the field, and they predict that the intelligent building market will expand to cover not only offices, but also other types of buildings, such as hotels and hospitals.

The diversified and sophisticated needs of customers encompass all construction, from factory and office buildings to residential homes. *Century-lasting houses* are those homes that builders claim will endure for a 100 years. These homes are not only structurally solid, but also allow for easy changes in the room layout and an easy replacement of household equipment and fixtures in accordance with changes in family membership and life-styles. Condominiums officially recognized as a century-lasting housing system by an authorized certification organization are now selling strongly.

A demand is growing for houses designed specifically for elderly widows or widowers wherein they can live alone in safety and comfort. Home builders are also promoting homes fully furnished with automatic equipment. On the other hand, elegant "handmade" houses are popular among a particular segment of home purchasers. Because houses were in short supply until about two decades ago, Japanese people were happy to live in look-alike government housing, but because housing units outnumber households today, people are not satisfied with what they consider to be low-quality housing. The main current in the construction market overall is moving from quantity to quality satisfaction, forcing contractors to upgrade the quality and sophistication of their products, be they factories, offices, stores, or houses.

Tougher Competition

Customers are getting both choosy and fussy, but market growth has stopped. The prospects are slim that an unfriendly business environment will improve to any appreciable extent in the foreseeable future. Contractors are forced into ever fiercer competition. In November 1983, a cluster of brand-new buildings, collectively known as Ogikubo Town Seven, was opened as the centerpiece of an urban redevelopment project in Ogikubo, a mainly residential community in

Tokyo's western suburbs. These plush, ivory-colored commercial buildings in front of Ogikubo Train Station composed the first major urban redevelopment project in Tokyo, which involved 101 local landowners who had been persuaded to combine their properties as part owners of the entire redevelopment premises. Shimizu Corporation managed this nine-year project as the main consultant, planner, designer, and constructor.

Urban redevelopment projects such as Ogikubo Town Seven are considered to be one of the most promising markets for contractors. Although public administration offices hitherto have taken the initiative in such projects, partly because of today's tighter budgets, government offices encourage private enterprises to take the lead. But the Ogikubo Town Seven is a good example of just how much time and human energy these community-based projects take for completion. During the high-growth economic period, contractors were able to obtain orders, especially private orders, almost effortlessly. Large general contractors got 80% of orders without having to compete with rival bids, because long-standing business connections with clients were respected. Accordingly, the general contractors were able to carry out the planning, designing, construction, and delivery with a silky smoothness.

Recently, however, the sellers' market has become a buyers' market, and once-loyal clients now place economy before long-standing connections. Today the general contractors acquire less than 60% of new orders without bidding. Construction projects that require a vast amount of time and labor for planning and coordination are increasing because an increasing number of projects demand that the contractors act as the go-between for landowners and other interested parties before a project can be launched.

Heated Competition with Outsiders

Competition is widespread, and the competitors are not necessarily contractors. Large makers of electrical machinery, steel products, and ships are now active in the construction market, and some rank high as contractors (see Table 1.4). Although currently their construction work is centered around equipment installation, the number of competitors from outside the construction industry is certain to increase. With the growing need for automated equipment in factories, offices, and homes, the borderline between the construction of structures is gradually disappearing, and the interface between the construction and nonconstruction industries is growing correspondingly wider, exposing contractors to competition from a greater number of outside companies.

In the housing business, which provides a massive market worth 20 trillion yen ($130 billion) a year, a number of textile, steel, and electric appliance makers have set up home-building operations, taking advantage of specialties such as consumer electronics, appliances, interior fabrics, and other home accessories.

Table 1.4 Work Completion Ranking of Nonconstruction Companies (Unit: million dollars)

		Completions		
Ranking	Company	Construction Work	Installation Work	Total
7	Mitsubishi Heavy Industries	48	2,631	2,679
9	Ishikawajima-Harima Heavy Industries	125	2,213	2,338
25	Fujitsu	0	1,503	1,503
27	Nippon Kokan	656	721	1,377
32	Hitachi Zosen	7	1,188	1,195
36	Toshiba	0	947	947
41	Mitsui & Co.	234	607	841

For some time now, Toyota Motor has forged ahead as a house builder and a bed manufacturer, with a long-term view to developing housing operations as a major business line. Especially in the area of automated housing, makers of electrical machinery are pushing ahead with comprehensive R&D activities. Since information and communications equipment is expected to become an inseparable part of these automated homes, electrical machinery makers may become the leading contractors in this particular construction market. In fact, in all probability, competition with non-contractors will intensify in most areas of the construction market, including houses, factories, and office buildings, plus the prospective fields of energy-related, marine, and space development projects.

Predictable competition from foreign contractors should not be overlooked. Japanese contractors are already competing with them in third-country markets, but engagement on the domestic front is obviously foreseeable in the near future. The Japanese Construction Ministry states that no legal barriers exist against foreign construction firms bidding for business in Japan, but foreign contractors claim the contrary. The foreign contractors' complaints may stem from their difficulties in building close business relationships with Japanese subcontractors and obtaining a full knowledge of Japanese business practices and technical construction codes. Regardless of the reason, virtually no foreign contractor has obtained construction work in Japan to date, whereas Japanese contractors now garner about one trillion yen ($6.7 billion) worth of orders from abroad.

Foreign interest in the Japanese market has recently swelled; the Japanese market has an array of big projects on the agenda as part of the nation's domestic demand stimulation effort. The interest has become particularly keen since the construction booms in the Middle East, Southeast Asia, and elsewhere in the world came to an end. In the spring of 1986, the U.S. government filed a request for participation by U.S. contractors in the New Kansai International Airport project off the shore of Osaka, and soon similar requests were lodged by South Korea and European countries. In January 1987, the joint venture formed by six Japanese consultant companies entered a contract with Bechtel Civil of the United States for the basic planning of the New Kansai International Airport terminals. This agreement is considered to mark the beginning of full-scale foreign participation in Japanese construction projects. Participation by companies from the United States, South Korea, and Europe is expected for other big projects, such as the planned construction of overhanging highways across the Tokyo and Ise Bays, and Japanese contractors must be prepared for tough competition with outside contractors.

THE SEARCH FOR A NEW PATH

The president of Shimizu Corporation, Teruzo Yoshino, speaking recently before a group of first-year employees said, "The social environment is changing fast, and I believe this period of renovation will be a period of advancement for our company. Our Shimizu employees are not afraid of making mistakes but have courage, and I believe they are building a bright future through an entrepreneurial spirit." President Shoichi Kajima of the Kajima Corporation welcomed new workers by saying "Our industry is often regarded as frozen in an 'Ice Age,' but in my opinion, the future is not grim because we know there is a huge potential need." Chairman Hajime Sako of the Taisei Corporation made this statement before his company's new employees: "In comparison with other industries, our construction industry is undergoing a profound transformation, and which companies are the more capable and resourceful will soon become clear. Our company faces the challenge of evolving into a knowledge-intensive construction firm."

These serious messages by top executives reflect the trying environment in which the construction industry is placed: The market expansion has stopped. Customer needs are becoming ever more varied and refined, and thus more elusive. The business competition is bringing nonconstruction firms and foreign contractors into the same ring once occupied only by traditional foes. The market does not offer easy profits, and these executives are calling strongly for a spirit of undaunted challenge to overcome these difficulties. The nation's industrial structure is changing, and contractors incapable of adapting to these changes will be forced out of the market. The executives suggest that it will be possible for alert companies to find and develop new business opportunities and take a major step upward.

The market is no longer expanding, but still offers 50 trillion yen ($330 billion) worth of construction business every year, and although it looks stagnant macroscopically, some segments of the market are growing rapidly. Furthermore, many business opportunities can be found in construction border areas interlaced with various nonconstruction markets. Whether these business opportunities can be taken advantage of will depend on a company's capability to grasp changes in social environments and construction needs and to formulate long-term strategies to respond to these changes. In this sense, future strategies are of critical importance in today's Japanese construction industry. In the next chapter, the fundamental conditions to be considered in formulating future growth strategies for construction companies will be discussed.

Tokyo Electrical Power Sodegaura Thermoelectric Power Station LNG Storage Tank (Japan) Shimizu Corporation

CHAPTER 2

FUNDAMENTAL CONDITIONS FOR GROWTH

THE NEED TO FORMULATE STRATEGIES

In the process of carrying on corporate activities, companies are said to make three different types of decisions—strategic decisions, tactical decisions, and operational decisions. In Japanese companies, tactical and operational decisions are made in the course of daily activities, either by the manager concerned or at regular staff meetings. These decisions concern mostly familiar, predictable events and operations. On the other hand, if the corporation environment undergoes an extensive change, the company develops an urgent need for strategic decisions to enable the company to quickly cope with these changes in the surroundings, predict future changes in the structure of the environment, and establish new, appropriate relationships with the environment.

Consequently, the present need of the Japanese construction industry for strategic decisions is greater than ever. Today, the demand for construction work is changing dramatically in terms of both quantity and quality, and the competition is intensifying. Although contractors hitherto grew smoothly in an expanding market by simply making tactical and operational decisions, today their success or failure depends heavily on the capability to adapt to changes in the market environment. Thus, there is an increasing need for contractors to formulate fundamental policies for achieving growth. When trying to map out basic strategies, a company must accurately assess the conditions of the market and competition as well as its own position in the market. On the basis of these analytical efforts, it will become possible for companies to determine their goals and to plan strategies for reaching these goals.

Analytical Axes for Market Segmentation

The market in which contractors operate can be defined in terms of three axes (see Figure 2.1). On the first axis, which we imagine as extending perpendicularly from the plane of the page, is the product parameter. This parameter is bisected into the construction and civil engineering categories, and the construction category is further divided into a number of subcategories such as residential housing, office buildings, and production facilities. The civil engineering category includes site renovation, road, and sewage subcategories. It is possible to further divide these subcategories by setting customer or technology parameters. For example, residential houses may be divided into wooden, single-unit homes and high-rise condominiums. Production facilities may be grouped into general factories and high-tech, precision facilities such as clean rooms.

As a result of market expansion in the prosperous period, the range of marketable construction products has broadened to cover marine and underground structures,

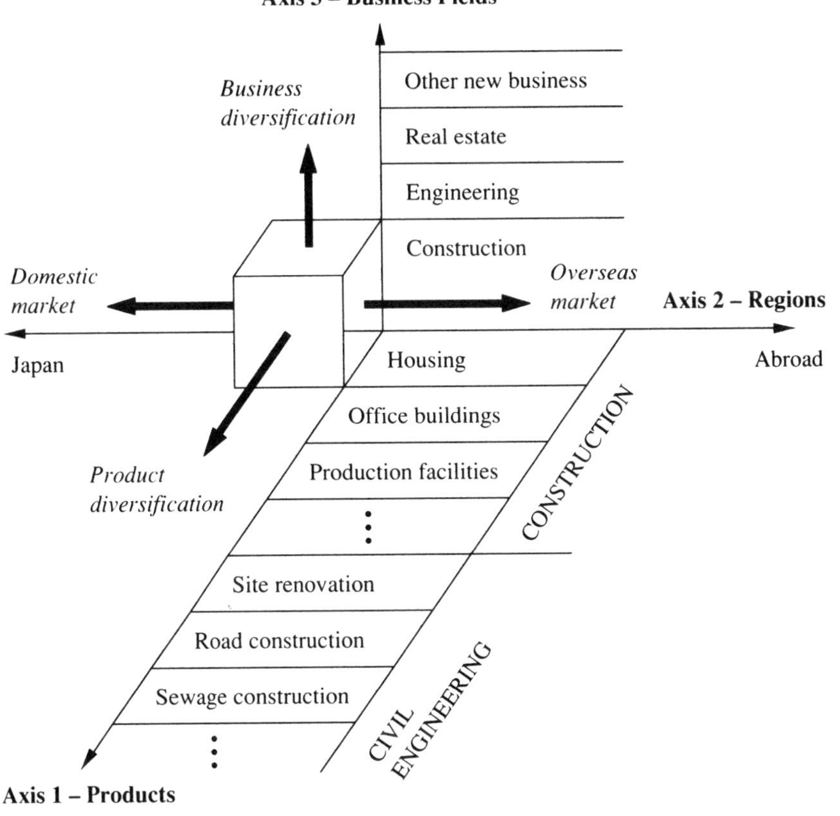

Figure 2.1 Three axes for market analysis

and many contractors have tried to expand their construction product lines to cash in on this expanding market. This *product diversification* along the first axis has been a popular strategy practiced by most contractors.

The second axis, which is imagined to be horizontal to the page, indicates the geographical locations of markets. Limited market areas such as villages, towns, and cities on the left side of the axis may be expanded to international markets on the far right, and each contractor must decide the exact part of the axis the company should target. Large contractors have striven to expand internationally, but it is equally feasible for construction firms to focus operations on specific regions in Japan in a bid to expand their share to the maximum within the specific

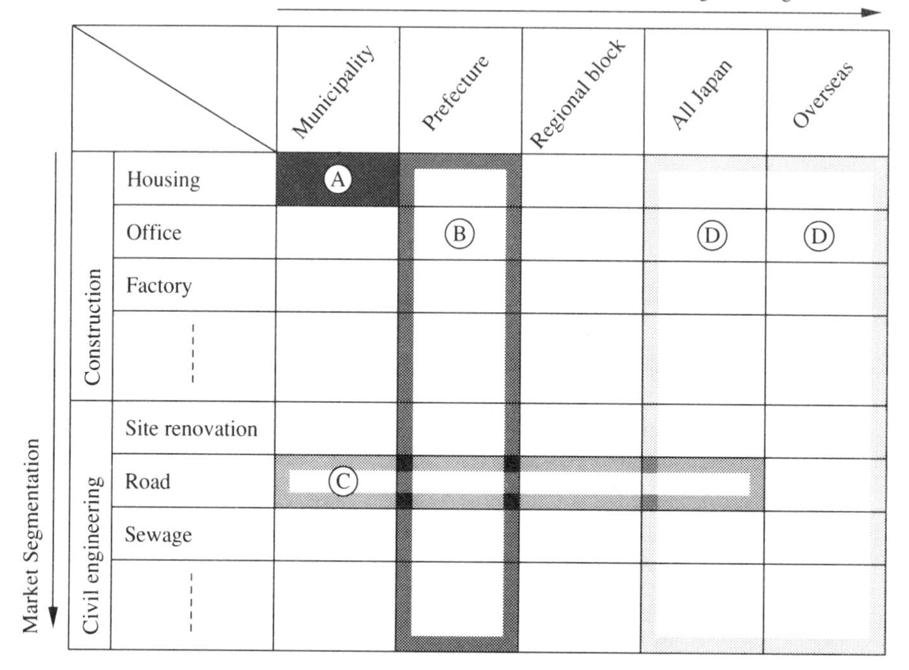

Figure 2.2 *Example of axis combination (axes and 1 and 2) for market analysis*

regional market. Large local contractors* compose the most successful category of companies that have stuck to the localization strategy.

The combination of the first and second axes provides for many areas of construction operations (see Figure 2.2). For example, Box A in Figure 2.2 shows the small local home builders operating on a town or city level, and Box B focuses on the leading local contractors operating in most construction and civil engineering fields on a prefectural (county) level. But due to the market stagnation of the past few years, the companies falling into Boxes A and B have been faced with the

* Large local contractors are top construction firms usually operating within a single prefecture. (Japan is divided into 47 prefectures, or county-like self-governing entities, having their own respective governments.) These contractors post annual sales of around 10 billion yen ($67 million) and command a close link with local banks and corporations. They often come into direct competition with nationally operating general contractors for locally based projects.

need to choose options. For example, they may have to expand the geographical areas of their operations, and if they decide not to expand geographically, then they will have to move into new segments in the existing market.

Box C indicates companies that specialize in specific construction or civil engineering fields—for example, road construction—but these companies operate nationwide. In their case, the options may be to expand their share in the existing market or branch out into new product lines. Shimizu Corporation, Taisei, Kajima, and other general contractors belong to Box D, operating in all market fields and all regions in Japan and also in overseas markets.

The third axis extends into the direction of business diversification, beginning with construction-related businesses such as real estate, engineering, and architectural design and consulting. Shimizu Corporation, for one, has expanded into the home-building business, engaging in a number of extensive new town development projects in different regions and building condominium homes for sale. In Shin-Matsudo, a brand-new commuter town in Tokyo's eastern outskirts, Shimizu has developed a total of 2600 high-rise condominium units on 160,000 square meters of land since 1977, a project which is still continuing. Kajima and Kumagai Gumi are two other general contractors now developing new commuter towns near Yokohama and Tokyo, respectively. Most of the big general contractors and some medium-sized contractors are forging ahead, with considerable success, with development projects involving the purchase of land—not only the construction of buildings and houses. As another move to expand their construction business peripheries, general contractors are eager to bolster their EC (Engineering Constructor) capabilities in a bid to expand their engineering and consulting businesses.

Business diversification attempts by contractors are also directed toward areas not closely linked with construction, but often in the sports, leisure, information, leasing, and other service areas, which in Figure 2.1 appear along the third axis. The objectives of these moves to explore new business arenas, related or unrelated to the construction market, may be summarized into three points.

1. Acquisition of improved engineering and design capabilities which will enable the contractor to upgrade products and services to meet the increasingly sophisticated needs of the construction market.
2. Stabilization of the contractor's profit by creating new profit sources outside of the construction market proper.
3. Creation of new jobs for older workers who may have to be shifted from the main construction operation in the future. In construction companies, as in most other Japanese corporations, the largest percentage of the employees are between the ages of 35 and 40, the people born during the postwar baby

boom. In 10 to 20 years these older workers are expected to pass their productivity peak but will have to be kept on the payroll under the widespread lifelong employment arrangement.

Criteria for Formulating Strategies

At what market areas should the contractor aim? How should these areas be approached? To plan growth strategies, answers to these questions must be determined by a comprehensive evaluation of the company's business potential on the basis of three evaluation criteria:

1. Market characteristics.
2. Status of competitors.
3. The company's resources.

This evaluation process will make it possible to determine whether the company should continue to pour its resources into existing markets, or whether an attempt should be made to enter a new market.

The first criterion requires the company to study market potential, market scale, and other market factors. Here, it is important to make a clear distinction between market size and market growth potential. Some markets may have great growth potential but the ultimate market size may not be large. In contrast, some markets can be large but the growth rates may be modest. The acquisition of a new home-building market worth 0.1 billion yen a year can be significant for a small-sized contractor of office buildings with annual sales of 2 or 3 billion yen, but for a larger contractor of high-rise buildings with annual sales of 100 billion yen, a new housing market worth only 0.1 billion yen is not worth consideration.

The concept of *market style* may be considered as a subcategory of market characteristics; it has to do with such questions as whether the market involves most of the customer segments or only a limited number of special customers, whether each lot of orders is large or small, and whether demand is geographically concentrated or dispersed. The house rehabilitation market is commanding increasing attention today, but contractors must keep in mind that the size of each rehabilitation order is small and that orders are geographically widespread and irregularly timed. As a result, the rehabilitation market is clearly different from the construction market in style, and to start a rehabilitation business, the contractor will be required to readjust the company organization.

For the second evaluation criterion, competitors, it is important for the contractors to keep tabs on the number and market share of their competitors, predict which nonconstruction companies are likely to move into the construction market, and compare their own strengths and weaknesses with those of existing and potential competitors.

The third criterion, company status, is the hardest item to evaluate accurately. Either a self-overestimation or a self underestimation can be dangerous. In this case, *self* means the business resources owned by the company, such as personnel, equipment and materials, money, information, and branch office networks. Technology and business experience are subfactors under the information and personnel factors. Limitations in personnel and monetary resources often prevent contractors from entering a new market. For instance, it would be extremely difficult for a medium-sized company to start a nuclear power facility construction business, because of its differential from large general contractors in technologies, experience, information backlog, and finance concerning nuclear power plants. Nevertheless, it is very possible that the smaller company has a greater number of workers more knowledgeable regarding, for example, the construction of a special type of condominium building in a specific region than workers employed by large general contractors. For this type of smaller contractor, the task is not only to keep a competitive edge but also to strive for an expansion of its share of the existing specialized market.

In planning to diversify into the realty and architectural consulting businesses, it is necessary to assess the availability of the company's business resources, including the possible extent to which existing branches and offices can be mobilized for new business. In addition, the purchasing potential of major customers must be studied, and clients with a large potential for future business should be protected from the approaches of competitors.

To accurately evaluate its own capabilities, each construction company may compare itself with competitors with regard to performance in each specific market segment. This enables the company to rate itself as strong, weak, or equal in each market. It will then become possible to obtain a clear view of market characteristics, competitors, and self-capabilities, and to predict a market share for the short-term future and a growth potential for the long-term future. As previously noted, the company's potential can be expanded by launching effective financial and R&D strategies. Although an evaluation may give only equal or weak marks compared to competitors, these marks can be improved to strong by mobilizing appropriate financial and R&D strategies, two key ingredients for corporate growth.

Strategic options can be more precisely defined by segmenting the market on the basis of the three-axis analysis, and the company's potential for each market segment is determined by the three-criteria evaluation. As a result, it will become possible for all contractors—large, medium, or small—to formulate a growth strategy leading to an emergence from the Ice Age.

GENERALIZATION VERSUS SPECIALIZATION

Although corporate growth strategies vary for each company, they may be divided into two approaches. The first is *generalization*, the approach most large contractors in Japan have taken, and the second is *specialization*.* In trying to formulate a growth strategy, a company must decide on the basic stance to be taken between the generalization and specialization approaches.

Generalization Strategies

Most general contractors in Japan have tried to grow by expanding the areas of construction operation and order-taking activities. As a result, leading general contractors, such as Shimizu Corporation, Taisei, and Kajima, operate in the entire range of construction and civil engineering fields from residential houses, office buildings, atomic power facilities, and underground structures to marine development projects. Today, general contractors are further expanding their operations into the engineering and urban redevelopment fields.

Although the generalization approach worked well for the general contractors during the period of rapid market expansion, the drawbacks of this approach are now becoming obvious. Generalization has caused all general contractors to operate in a similar way, and the Japanese construction industry as a whole has become very much a look-alike business community. Featureless generalization is uninteresting to customers, and the indistinguishability of the individual general contractor from many other contractors in the minds of customers invites stiff underpricing competition. Thus, large construction companies are confronted with the question of how far they can or should push ahead with their generalization strategies.

Today, the content of construction needs has changed. One of the changes is an increase in demand for high-tech oriented construction work, best exemplified by the building of clean rooms for high-tech industries. Another change is the increasing expectations of customers that contractors act as planners and coordinators in urban redevelopment projects, often involving the construction of large commercial buildings on a site owned by a large number of individual residents. Japan is considered to be still a relatively underdeveloped country in terms of social infrastructure, and its social needs are predicted to become more diverse on

* *Specialization* in this context means to pursue a special area, or niche, of the construction business, regardless of whether the company operates in other areas of the construction market. In electronic machinery production, Sony Corporation may be considered as a successful *specialized* company compared with all-around electronic machinery giants such as Matsushita, Hitachi, and Toshiba. Sony specializes in consumer audio-visual products, but its bigger competitors produce the entire range of consumer electronics products, plus office and industrial machinery.

both construction work and non-construction work such as planning and coordination. This will require contractors to upgrade and broaden their social development know-how.

Although general contractors operate over the entire range of the construction market, it will be impossible for any contractor to increase its financial and human resources to meet all the diversifying market needs. Indeed, it will be difficult enough for large contractors to remain good all-around companies in the construction market. Accordingly, the general direction large contractors may take is now gradually being revealed; the course they will follow will be toward the building of strong specialized business lines. This will compel each company to allocate limited resources to selective, advantageous business areas. General contractors will be required to pursue this type of specialization venture, while continuing to operate in general areas to keep business opportunities at the maximum level. Within the framework of generalization, for instance, general contractors may choose to bolster their bases of regional operation. Kumagai Gumi has grown into one of Japan's Big 6 general contractors by placing special business emphasis on international operations while also operating in most market areas.

This type of generalization strategy, with an eye to building a specialized, competitive line of business, will be even more important for both large (other than Big 6) and medium-sized general contractors.* Squeezed between the big general contractors and regional contractors, a medium-sized general contractor must build specialized strongholds of its own to survive, or face the danger of becoming a colorless company with unimpressive all-around features. To operate along the generalization line and at the same time build a specialized area appears to be contradictory, but this dual strategy seems to be the most promising way for Japanese general contractors to succeed in the homogeneous Japanese construction industry.

The Specialization Approach

Hasegawa Komuten Company, a large general contractor, is well known for the construction and sale of high-rise condominiums, and Hasegawa and its group companies handle all business associated with high-rise condominiums, including real estate, construction planning, architectural designing, and the after-sale main-

* There is a large number of medium-sized construction companies that operate extensively in Japan. Their annual sales range from $60 million to $600 million, and they capitalize at somewhere around $2 million to $30 million. These contractors clearly lag behind the large general contractors in technological, human, and financial resources, and furthermore, lack the strong local bases which leading regional contractors have consolidated to perform detailed marketing activities in their specific regional markets. Consequently, the medium-sized general contractors are subjected to greater pressure from a construction slump.

tenance and management of condominium buildings. Although its business performance has slackened a little over the past two or three years because of a general market slowdown, Hasegawa still boasts a high profit margin envied by other general contractors. Hasegawa's ratio of current profit to sales reached a 12 or 13% level at one time, and registered a reasonable 5.4% for the term ended May 1986. The company is reportedly planning to expand into the U.S. market in its strongest field—the construction of condominium buildings. Unlike most other general contractors, Hasegawa specifically developed the condominium market for the general homeowning public, combining a realty business with the construction operation. This has enabled the company to enjoy an extraordinary profit margin in the price-competitive construction market.

Several other companies in the construction industry have become successful contractors through a specialization approach. Kandenko Company has been successful with an electricity facility engineering operation, and Nippon Hodo Corporation and Nippon Road Corporation are two leaders in the specialized road-paving business. For the term ended March 1986, Kandenko registered sales of 300 billion yen ($2 billion) and a current profit of 15 billion yen ($100 million); corresponding sales were 20 billion yen ($130 million) for Nippon Hodo and 95 billion yen ($630 million) for Nippon Road.

The specialization approach nevertheless embodies a unique risk—specialized companies can easily stumble if the market stops growing, whereas generalized firms have greater resistance because of their diverse sources of income. An example of this phenomenon is the road-paving industry, which enjoyed brisk growth in public road investments in the past but has seen this growth reach a plateau for three or four years in the early 1980s (see Figure 2.3 and Table 2.1). Mainly because of the government's fiscal squeeze, the Japanese road-paving market is not likely to enter another period of growth in the foreseeable future. Consequently, road-paving companies are now compelled to move into overseas markets or branch into nonpavement markets. Leading paving companies are particularly interested in the construction business, such as the construction

Table 2.1 Road and Highway Development in Japan

As of end of fiscal year:	1960	1965	1970	1975	1980	1983
Length of expressways (km)	—	190	649	1,888	2,860	3,435
Length of urban highways (km)	—	39	164	199	286	310
Rate of national and prefectural road improvement (%)	29.0	38.9	52.9	62.2	67.3	69.0
Rate of municipal road improvement (%)	1.5	4.4	12.0	27.0	41.3	46.8

Source: 1984 White Paper on Construction, Tokyo: Ministry of Construction, 1984.

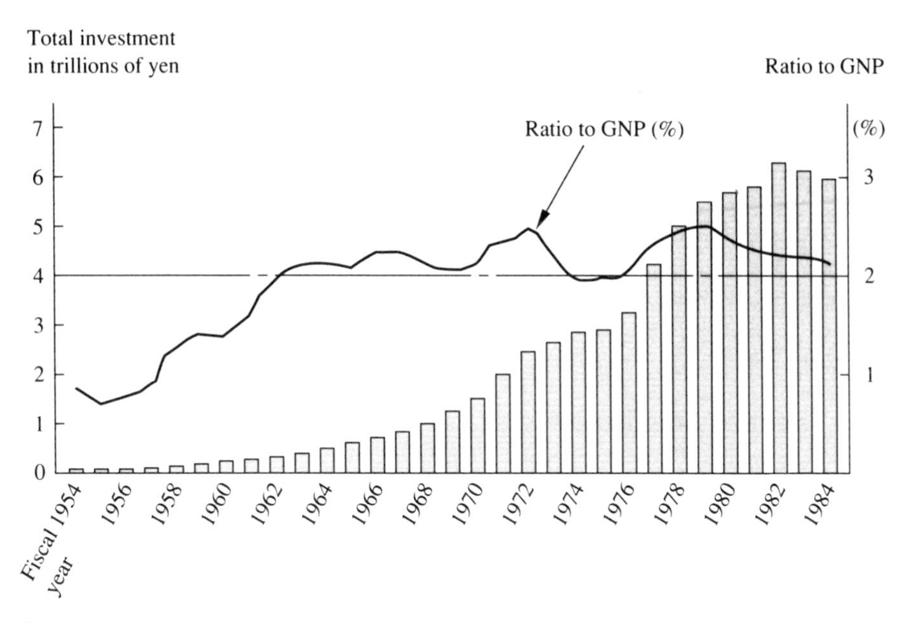

Source: Monthly of Construction Statistics (Kensetsu To'okei Geppo'o), Ministry of Construction, and Research Council for Prices in the Construction Industry, January 1985.

Figure 2.3 Japan's total road construction investment

of tennis courts and other sports facilities. Some have already set up construction subsidiary companies and are eagerly recruiting experienced construction engineers.

Marine civil engineering companies, which are engaged mainly in reclamation and dredging work, experienced diminished market growth at an earlier stage than the road-paving companies. Orders have remained in a slump since the first oil crisis in 1973, as the pace of harbor construction slowed and as the number of petrochemical, steel, and other energy-guzzling plant complexes, which were usually built on reclaimed coastal sites, decreased sharply. This has forced marine civil engineering companies to enter the land civil engineering and construction markets and start overseas business operations. These companies are also trying to develop marine technologies of their own, to meet a predicted increase in the demand for marine development projects in the future.

Clearly, at least two things must be done before deciding on a specialization strategy.

1. Assess the current growth of the market in question and its future growth potential.

2. Consider the possibility of an easy advance into other markets in case the market in question is nearing the end of a growth period.

In this connection, the rehabilitation work market in Japan is regarded as one specialized area having a large potential for growth. Britain serves as a good model when trying to predict Japan's future rehabilitation market (see Figure 2.4). In Britain, there are many buildings surrounded by scaffolding; they are not under construction but are undergoing exterior and interior repair work. In recent years, the rehabilitation business has expanded markedly and now accounts for nearly half of the British construction market. Specialized rehabilitation companies have increased accordingly in both number and individual scale.

THE JAPANESE REHABILITATION MARKET. Japanese contractors began to pay a great deal of attention to the rehabilitation market from the early 1980s, mainly in the housing area. As the background to this development, Japan suffered from a serious housing shortage for many years after the end of World War II, and the Japanese government was compelled to place an increase in the number of housing units as the top housing priority. Accordingly, the number of housing units grew sharply, exceeding at last the number of households in all Japanese regions in 1973 and prompting the government to shift the policy emphasis from a housing unit increase to quality improvement—toward more spacious and comfortable homes.

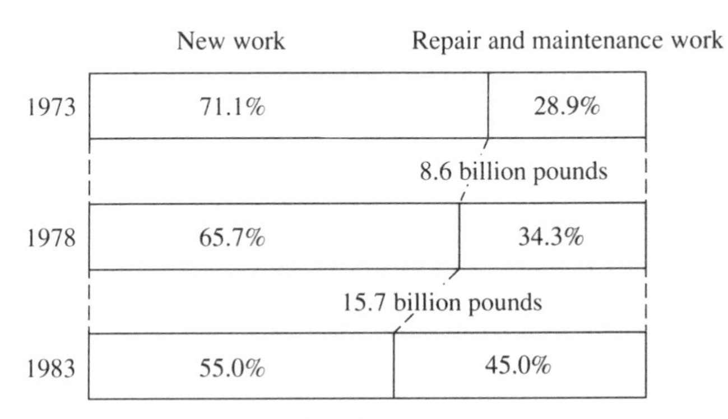

Source: Produced from *Housing and Construction Statistics 1973-83,* U.K. Government Statistical Service; *Construction Industries of Major Western Countries,* Tokyo: Research Institute of Construction and Economy 1985.

Figure 2.4 Construction work in Britain

Reflecting these developments, home builders and home equipment makers have launched full-scale promotions of home rehabilitation and interior improvement business.

Home rehabilitation has revolved around single-family houses so far, but in the future, the rehabilitation of collective housing units, such as apartment houses and condominiums, is expected to increase in Tokyo, Osaka, and other big city areas.* The residents of collective housing units in big cities are increasingly keen to renew the exteriors, interiors and household fixtures of their homes, and home rehabilitators also anticipate an increase in rehabilitation demands due to a nationwide home automation movement brought about by the advances in electronics technologies.

At present, however, large general contractors are more interested in the rehabilitation of offices, stores, hotels, and other commercial buildings, many of which were built in the 1960s and 1970s and are now due for rehabilitation. In addition to needs due to age, the contractors are betting on a brisk increase in the rehabilitation of office buildings to incorporate advanced information-communications equipment and office automation systems. Mitsubishi Estate, Japan's largest real estate company and the leading office building owner in central Tokyo's Marunouchi business district, is now making steady investments in the upgrading of the information-communications capacities of its rental buildings in the hope of collecting higher rental charges from tenants. Also, an increased demand for rehabilitation is anticipated from stores and hotels, which are more and more anxious to attract a greater number of customers by refurbishing their facilities.

In response, Shimizu Corporation, Tobishima Construction, and other general contractors have set up specialized rehabilitation subsidiaries to replace the former "maintenance" departments which, upon requests from previous clients, rehabilitated the buildings constructed by their companies. These specialized rehabilitation subsidiaries signal a change to an active rehabilitation marketing effort from the wait-for-a request operation, and reflect the sanguine views of most contractors on the future of the rehabilitation market.

In addition to housing units and commercial buildings, general contractors regard highways, railroads, bridges, dams, and other infrastructure projects as important prospective components of the rehabilitation market. Although rehabilitation has been outshone by the sustained increase of new infrastructure projects designed to elevate the low status of Japanese infrastructure to the Western level, the nation's social capital stock is unmistakably increasing, and the demand for a rehabilitation of infrastructure will certainly increase in the foreseeable future. In this connection,

*Tokyo and Osaka in particular have been hit by a cancerous rise in land prices, and the infamously low residential standards of the people living in these metropolitan areas have not been improved as yet.

the leaders of the construction companies and government offices are paying close attention to "America in Ruins," a book written by Pat Choate and Susan Walter in the United States, and are voicing the importance of investment in the maintenance and renewal of infrastructure facilities.

An estimate of the size of today's rehabilitation market in Japan may be possible from the long-term predictions of construction demand by the Ministry of Construction (see Table 1.2). According to the breakdown of the total demand provided by the Ministry, the rehabilitation market amounted to 2.6 trillion yen ($17 billion) in 1983 and will reach a predicted 5.1 to 5.4 trillion yen ($34–36 billion) in the year 2000, expanding at an average pace of 4.0% to 4.4% per year. This growth rate is considerably higher than those for private housing (2.4% to 3.4%), corporate construction projects (2.8% to 3.7%), and public infrastructure investments (−0.5% to 3.9%). The breakdown suggests that the rehabilitation market size as of 1988 will be roughly 3 trillion yen ($20 billion) a year.

The preservation of historical buildings and streets, an offshoot of the main rehabilitation market, is an important cultural issue in many countries. In Europe, stringent regulations are in force for the preservation of old castles, churches, buildings, and streets, but in Japan, such control has been lenient, with the exception of very famous temples, shrines, and the like, and many historical structures have been demolished as a result. Recently, however, citizens are more anxious to preserve old structures or move them intact to new locations, even when promoting a redevelopment plan for their community.

For example, in the face of a plan conceived by a group of private companies to revitalize Tokyo's central district around Tokyo Station, the Architectural Institute of Japan and other citizens' groups have petitioned for the preservation of the 70-year old Tokyo Station building, the largest brick structure in Japan. As these movements are expected to increase throughout Japan, general contractors, when managing various community development projects, are required to present appropriate construction plans that will enable the community to preserve its important structures and landmarks passed on through many generations.

MERGERS AND COLLABORATIONS

The Japanese construction industry comprises about 520,000 construction firms, from big companies with more than 10,000 employees each to one-man operations. It is impossible under present conditions for all of these contractors to grow. In their attempts to grow, or even to stay afloat, many companies now find that their technological, financial, and human resources are limited, and an increasing number will consider a merger or a business collaboration to deal with these limitations.

Trends toward Mergers

In early 1985, a construction trade newspaper printed two news items about companies formed by a merger of medium-sized construction firms. The first item was about Seiki Tokyu Company born from a union in 1982 between a road-paving company operating in the public road market and another ground-paving firm specializing in the private sector market. Through the merger, the two companies were able to expand their fields of operations and open new asphalt plants and paving work offices. As a result, the merged company became one of the largest companies in the paving business, and its annual sales expanded smoothly from a combined 53 billion yen ($350 million) by the two partners in the April 1983 term to 67 billion yen ($450 million) in the April 1986 term.

The second news item was about Daioh Construction Company, formed in 1972 through the merger of two large regional construction firms in Shikoku. One partner was active in land-based civil engineering, and the other operated primarily in harbor-related civil engineering. Due to the resultant expansion of business areas, the new company was able to grow despite the first oil crisis in 1973, and its sales grew more than fivefold, from 8.1 billion yen ($54 million) in the June 1973 term to 42 billion yen ($280 million) in the June 1986 term. Similarly, the company's capital rose from 240 million yen ($1.6 million) to 600 million yen ($4 million) during the same period. The merger has achieved the initial goal: to expand the two regionally based companies into a larger company operating on a national scale.

These two instances of merger by medium-sized construction companies demonstrate how a merger has brought about a noticeable improvement in the combined marketing capability of the merged company as a result of an expanded operational range. Similar mergers among medium-sized contractors will probably increase in the future, as this is a viable means to compete with the large general contractors on one hand and the regionally rooted small companies on the other. An industrial association of medium-sized contractors, the Japan Association of Representative General Contractors, recently released a research report in which it recommended that its members "study the possibility of merger while the company still has reasonable corporate strength."

The advantages of a merger may be summarized as follows. First, the partners supplement each other by combining their respective marketing areas, potential clientele, technological capabilities, and office networks. Second, it is possible to reduce total costs by eliminating overlapping expenses for office administration, research and development, and so on. A merger also enables the two parties to utilize subcontractors with economy and to lower the unit purchasing or leasing price of necessary equipment and materials, such as scaffolding units, through an increased purchasing or leasing volume. Third, because a merger expands the corporate scale, the new company's customers and the business community in general will think that it is more reliable. Consequently, the new company will have an improved position in business negotiations and experience a greater ease in producing business funds, which will make it possible to attract talented staff.

MERGERS IN THE WEST. Moreover, after considering mergers between medium-sized contractors, another question arises: Could there be mergers between large contractors and medium or small-sized contractors? In Western countries, mergers and acquisitions are more frequent than in Japan, and there are many examples of large contractors acquiring their smaller counterparts. Turner Construction of New York, one of the largest contractors in the United States, has acquired a number of regional construction companies, mostly in the South, since the early 1970s. Turner's primary aim was apparently to expand its operations into new regional markets without shifting the company's existing resources, but by utilizing the resources of the regionally established companies instead. Turner always selected companies that had non-union workers.

Trollope and Colls, based in London, is a medium-sized construction company that has been in operation for over two centuries, since 1778. The company posts annual sales of approximately one million pounds, mainly in the southeast region of England. In 1968, however, this firm was acquired by the Trafalgar House Public group, a conglomerate centered around the construction business. In retrospect the Trollope and Colls management believes that, by joining the group, the company was able to win strong support in both finance and management. In the British construction market, where growth in real terms stopped in the early 1970s, even the long-established Trollope and Colls found that the only way to survive was by sheltering under the wing of a conglomerate.

In Japan, corporate acquisition is usually a case of the absorption of the small by the large, with the small company ceasing to exist as a separate entity. But in Western countries the large often purchases the small as a subsidiary, thus allowing the small to continue operations under the separate corporate name but with the financial and managerial support of the parent company. These acquired companies are included in the parent's consolidated statement as subsidiaries or affiliates.

MERGERS IN JAPAN. Given the Western examples, there will likely be an increase in the purchase of or capital participation in smaller constructors by large general contractors in Japan. An increased number of small and medium constructors will be willing to receive support from their big brothers to bolster financial, technological, and project management capabilities. On the other hand, through mergers, the large general contractors will be able to expand their access to locally based construction projects without investing in additional business resources, such as personnel, goods, money, and information. Further, mergers could improve the consolidated performance of the large general contractors and increase job opportunities for their older employees, who may have to be relocated due to diminished productivity.

In 1986, there were two cases of mergers between large general contractors and locally based medium-sized constructors. A leading general contractor, Toda Construction Company (capitalized at 9.7 billion yen, or $65 million, and recording 358 billion yen, or $2.4 billion, in sales for the term ended September 1986), merged with and absorbed Shimato Construction Company (capitalized at 845 million yen, or $5.6 million, and posting 24.6 billion yen, or $160 million, in sales for the same term). Mitsubishi Bank, the main bank of both, acted as a go-between. Through this merger, Toda Construction hopes to inherit Shimato's clients and acquire new clients within the Mitsubishi corporate group market.

The other instance was the acquisition of a local construction firm, Asahi Construction Company (capitalized at 400 million yen, or $2.7 million, and posting 13.6 billion yen, or $91 million, in sales for the term ended March 1986) by a large general contractor, Wakachiku Construction Company (capitalized at 5 billion yen, or $33 million, and recording 73 billion yen, or $490 million, in sales for the same term). The larger Wakachiku is based in the southernmost part of Japan, Kyushu, and is most active in harbor construction and other civil engineering operations. Asahi was a construction-centered company in Chiba Prefecture, adjacent to Tokyo. Wakachiku intends to obtain a foothold in the Tokyo area construction market through the purchase of Asahi Construction.

Because mergers and acquisitions are relatively rare events in Japan, largely because of the tight-knit personal relationships between managers and employees, it is predicted that, in the initial stage, the majority of business collaborations will take the form of capital participation, in which the larger partner will purchase only a part of the smaller partner's stock. After this initial period, mergers and total acquisitions will gradually increase.

Increasing Collaboration with Other Industries

In December 1985, Shimizu Corporation set up a joint venture company, Beverly Japan, with Beverly Enterprises, the largest nursing service company in the United States. The joint venture has begun to launch planning, design, maintenance, and management services for old people's homes and other facilities for senior citizens. Shimizu's know-how in the construction of medical, welfare, and residential facilities and Beverly's special knowledge of the operation of facilities for senior citizens are now combined to initiate a new type of business in Japan, with the rapid aging of the population as the background.

The collaboration of construction companies with partners in nonconstruction industries is certain to increase in scope and diversity. Three basic types of collaboration are likely to develop. The first is the conventional collaboration in construction-related areas, involving mainly the research and development of construction technologies and products. Collaboration for the development of new construction materials belongs to this category. In the second type of collaboration, construction and nonconstruction companies work together to stimulate construction and related demand for their mutual benefit. For instance, contractors and trust banks work together to stimulate construction investment by land owners. The third type of business collaboration is aimed at launching businesses in nonconstruction fields. As in the case of Shimizu's partnership with Beverly, construction companies must team with specialists in orders fields to ensure a smooth entry into a brand-new business area. If both partners can combine their company resources and business know-how for mutual advantage, business collaboration will become a powerful means of growth for construction companies.

CORPORATE SHAPE-UP: INCREASING PRODUCTIVITY

It is an open secret that Japanese products are strong in the world marketplace because of good quality and high labor productivity, but productivity in the Japanese construction industry is exceptionally low compared to manufacturing companies. In an all-industry comparison made by a Japanese business consulting organization, Japan's manufacturing productivity surpassed the European level in

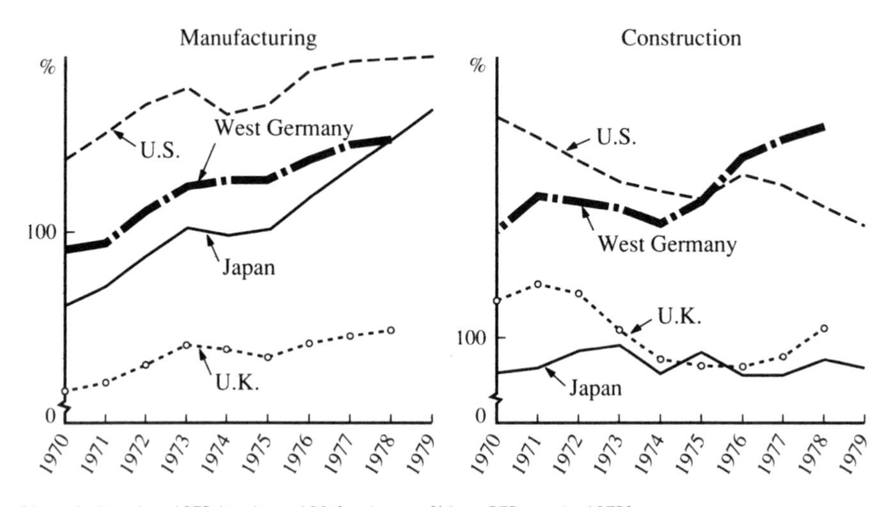

Note: Indexed at 1973 levels as 100 for Japan. [$1 = 273 yen in 1973].

Source: Study on International Comparison of Labor Productivity (Ro'odo'o Seisansei no Kokusai Hikaku ni Kansuru Kenyu'u), (Tokyo: Japan Productivity Center, 1982), 110.

Figure 2.5 Productivity in principal countries

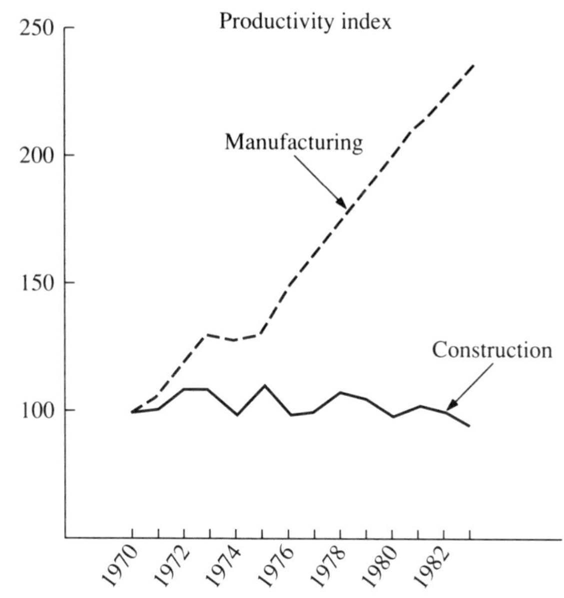

Note: Index on 1970 level as 100.

Source: Produced from *Annual Report on National Economic Computations (Kokumin Keiza Keisan Nenpo'o)*, Economic Planning Agency. *Labor Force Survey (Ro'odo'oryoku Cho'osa)*, Management and Coordination Agency.

Figure 2.6 Per worker productivity in Japan

the late 1970s, and is estimated to have exceeded the U.S. level in the early 1980s (Figure 2.5). Nevertheless, if the construction industry is compared, Japan remains far below the Western productivity level. What is worse, construction productivity in Japan has steadily declined over the years, while the average per-worker productivity of the consumer electronics, computer, and other manufacturing industries has climbed steeply (Figure 2.6).

Heavy Dependence on Manpower

Because construction is a labor-intensive service, it is impossible to map out growth strategies for contractors without grappling with the issue of the low productivity of Japanese construction. As a study by a research institute makes clear (Figure 2.7), the production increase achieved by the Japanese construction industry has always been closely linked with an increase in the labor force, and the total construction volume has ceased to grow since the mid-1970s largely because the number of

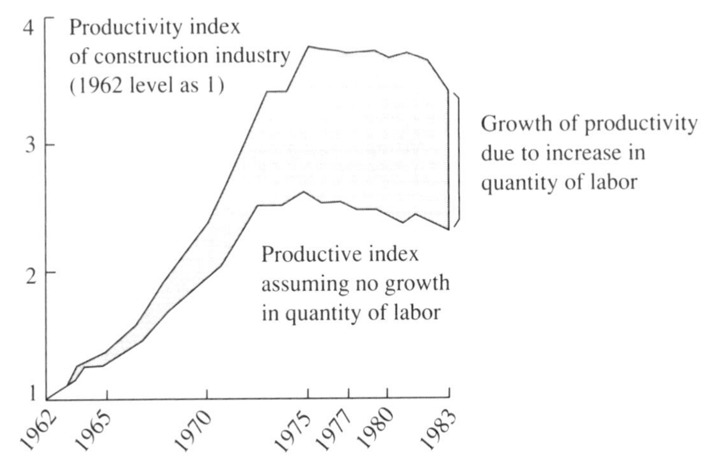

Source: Produced from *Analysis of Labor Demand Factors in the Construction Industry (Kensetsugyo'o ni okeru Ro'odo'o Juyo'o Yo'oin no Bunseki)*, (Tokyo: Mitsubishi Research Institute Inc., 1980).

Figure 2.7 Labor factor in Japanese construction productivity

construction workers has not increased. In addition, the productivity per worker in the construction industry has been slipping, partly because of the shortage of new skilled workers and partly because of a standstill in the progress of routine construction technologies.

Construction is often considered synonymous with "low wages and many accidents" by young job seekers, and contractors have failed to attract a sufficient number of talented young workers, causing the work force to age and become less skillful (see Figures 2.8, 2.9). The shortage of skilled labor often compels contractors to hire more unskilled workers just to complete the immediate work by the deadline, causing productivity to fall even further. Although the shortage of skilled workers has become less severe since 1980, this is simply because the volume of construction work has declined since that year (see Figure 2.10). If Japan's total construction investments expand only at an annual average rate of 2.9% up to the year 2000, as projected by the Research Institute of Construction and Economy (a think tank affiliated with the Construction Ministry), the shortage of skilled workers will continue to exist for the foreseeable future.

Another problem in the Japanese construction labor situation is that the necessary labor is not always available at the time when the industry has the greatest need for a larger labor force. In fact, the number of construction workers often increases when the industry is relatively well staffed, and often decreases when the industry needs more workers. This is partly because the labor available to the construction

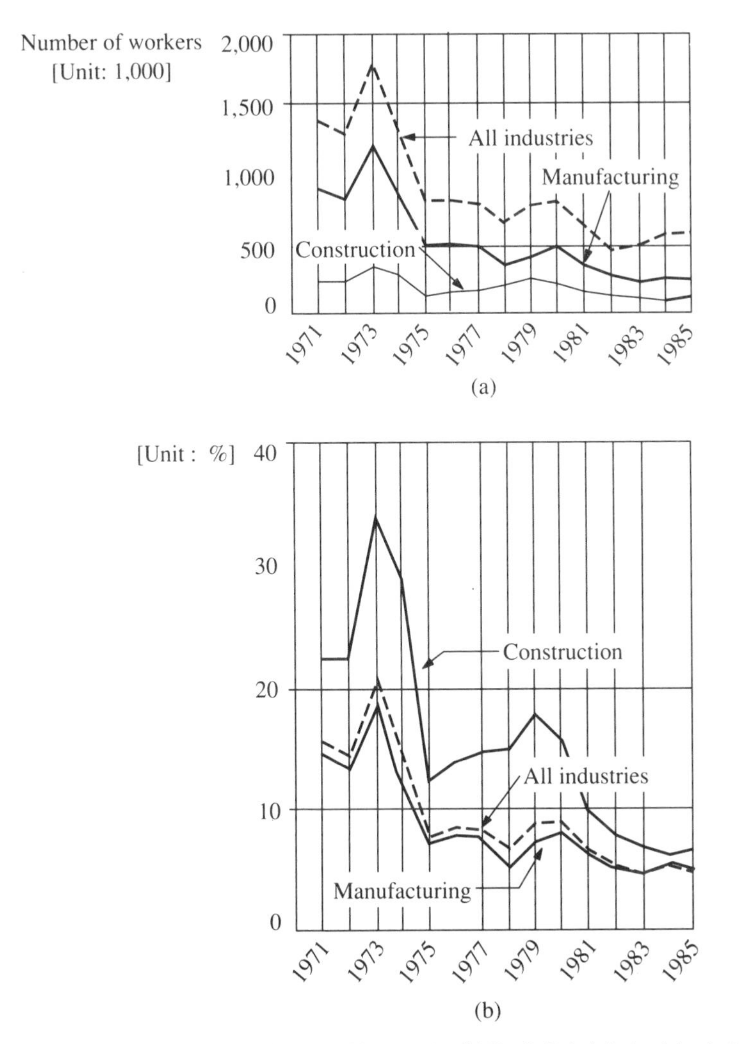

Source: "Survey Report on Demand and Supply of Skilled Labor" (*Gino'o Ro'odo'osha Jukyu'u Jokyo'o Cho'osa Ho'okokusho*), *Statistical Annal,* Tokyo: Ministry of Labor.

Figure 2.8 Skilled workers in the Japanese construction industry: (a) the short supply; (b) the declining percentage.

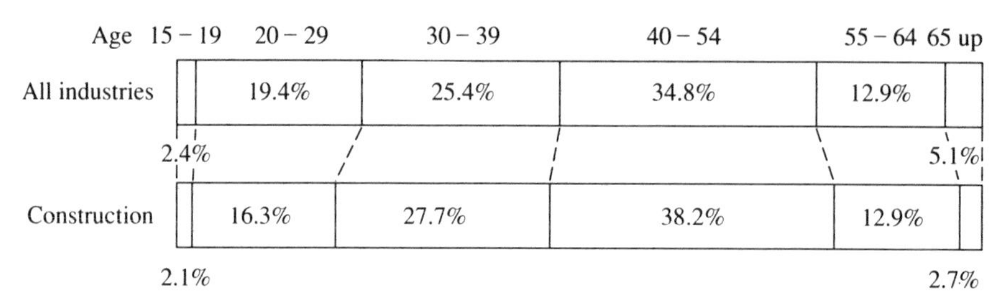

Source: Produced from *Labor Force Survey (Ro'odo'oryoku Cho'osa)*, Tokyo: Management and Coordination Agency.

Figure 2.9 Age of the Japanese work force in 1985

industry is strongly affected by the level of the seasonal labor influx from farming communities and the labor demand-supply conditions in other industries. If Japanese construction firms wish to improve their labor productivity, they must acquire a work force that is not affected by the manpower situations in other industries. Specifically, they must attract young workers away from other, more glamorous industries by increasing wages and fringe benefits and by upgrading working conditions. The main, if not the only, way for construction companies to be able to afford such incentives for workers is to improve their productivity, which is equivalent to increasing business profitability. Although this is a difficult question to solve (which comes first, the chicken or the egg), a breakthrough must be found if there is to be a brighter future.

Another cause of the drop in construction productivity is the standstill in routine construction technologies. As is evident from Figure 2.11, technology has made few contributions to the improvement of productivity in the construction industry—in sharp contrast to manufacturing industries. Although construction technologies have made remarkable advances in high-tech areas such as ultra–high-rise buildings and special-purpose structures, the majority of construction technologies that are used in most construction sites on a daily basis have not progressed for the past 10 years or so. The microelectronics revolution that has jacked up productivity in factories and offices has had little effect on construction work, one of the last fields where human muscle still plays a principal role.

Dispute with Residents

Note that, when planning and executing construction work, relations with the residents around the construction site can be a particularly large stumbling block against productivity in Japan. In this densely populated country, it has become increasingly difficult to undertake construction projects without eliciting hard feel-

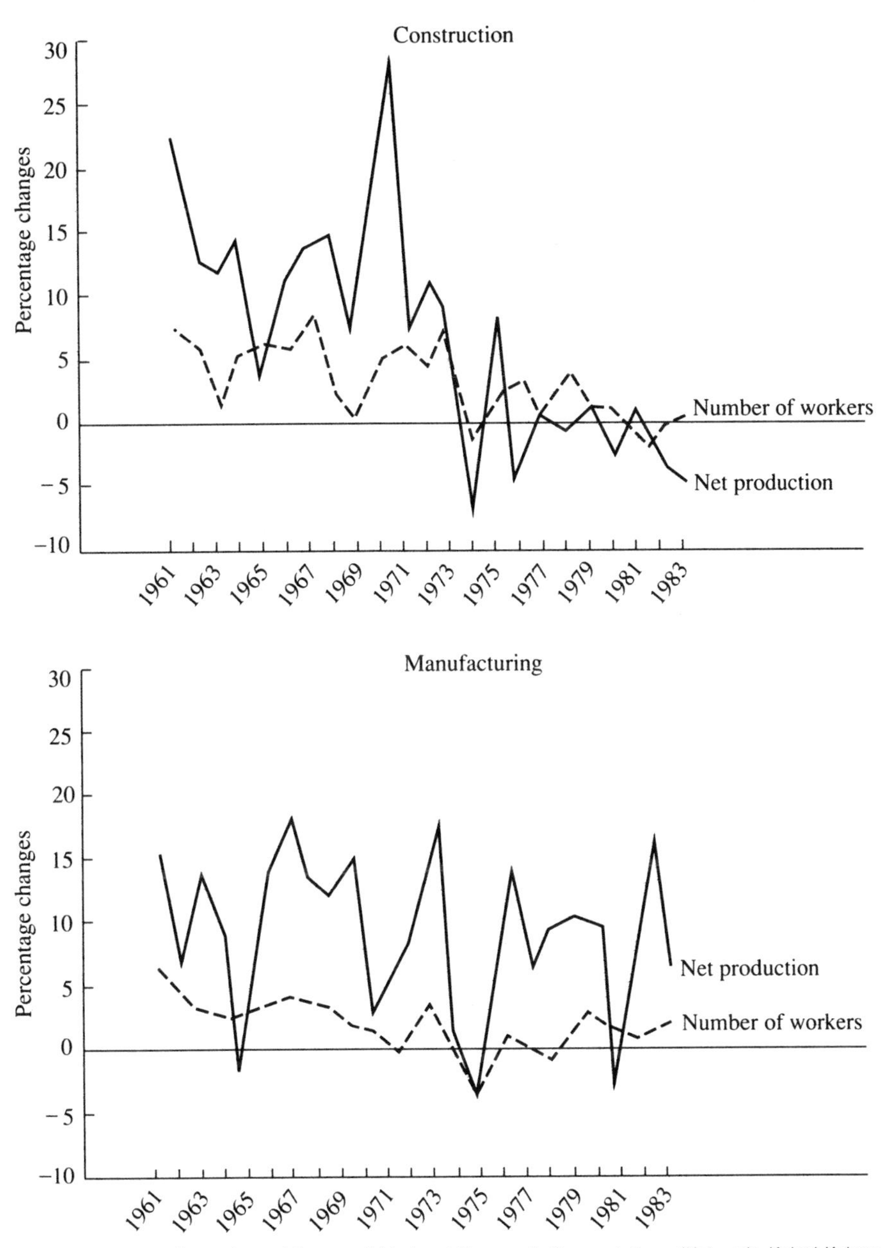

Source: Produced form *Annual Report of National Economic Computations (Kokumin Keizai Keisan Nenpo'o)*, Economic Planning Agency.

Figure 2.10 Changes in growth rates of net production and number of workers in Japan

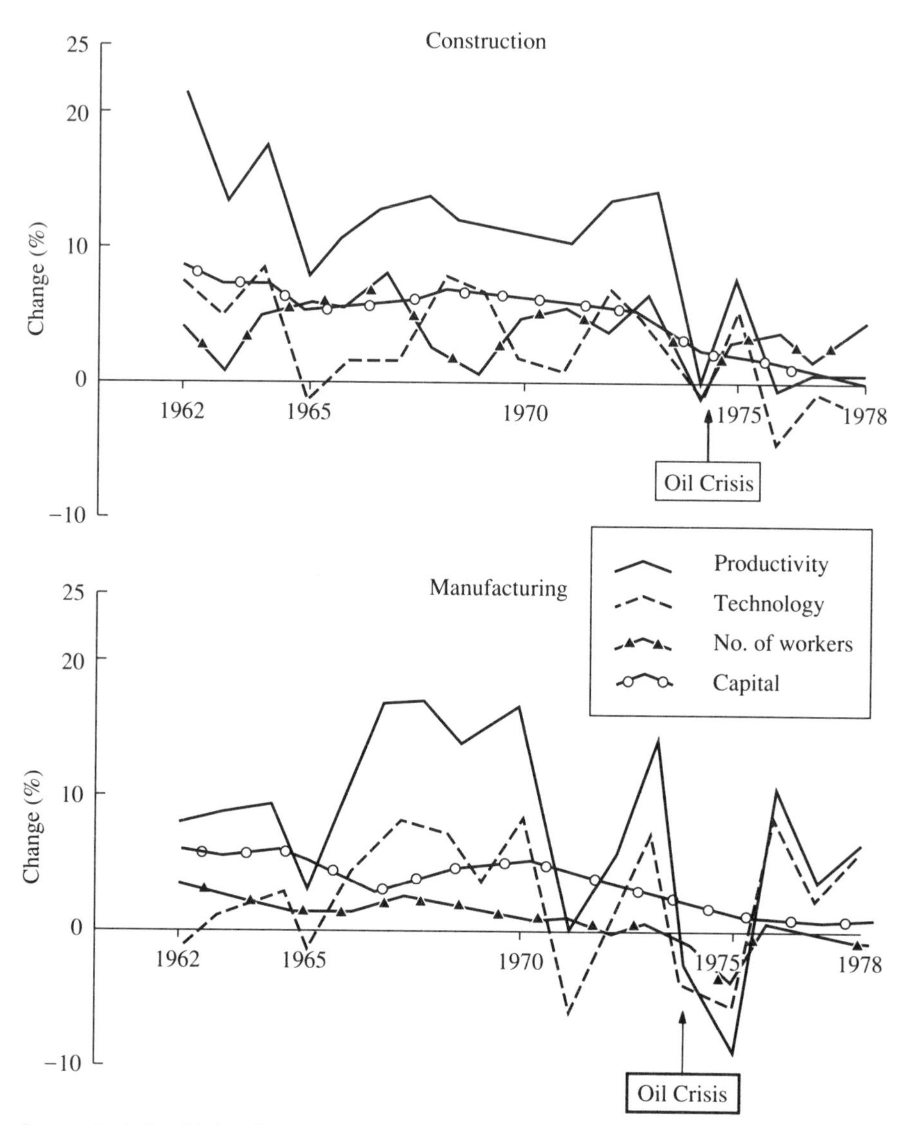

Source: Analysis of Labor Demand Factors in the Construction Industry (Kensetsugyo'o ni okeru Ro'odo'o Juyo'o Yo'oin no Bunseki), (Tokyo: Mitsubishi Research Institute Inc., 1980), 38.

Figure 2.11 Contributions to growth and productivity in Japan

ings from the neighbors about construction noise, blocked sunlight, and other environmental problems.* Accordingly, administrative offices have strengthened regulations on construction work plans, which caused, and still causes disputes to be prolonged and productivity to remain below its potential level. These disputes should be settled by the clients, but contractors have been often compelled to take a greater initiative in the settlement, to smooth the progress of projects. Thus, contractors are being given greater opportunities to utilize their professional know-how acquired as project organizers—the skill to manage project (disputes) and the skill to bring about a compromise among the parties involved (disputants).

Systematizing New Technologies

Using robot technologies, the automaker can manufacture different models of cars on the same production line; to varying degrees, manufacturers have benefited from microelectronics and robot technology in their efforts to improve productivity. These labor-saving technologies are only now appearing in the construction industry, and yet, to the surprise of many construction people, a Japanese university professor who attended the International Conference on Construction Robots held in Marseilles in June 1986 reported that Japanese construction robot technology appeared to be by far the most advanced in the world (Table 2.2).

The Development Project for High-Level Construction Technology Systems Using Electronics was launched in 1983 jointly by the government and private sectors, with the object of increasing the use of construction robots. Since automakers and other manufacturers took about 10 years to master robot utilization in their plants, it will probably take several more years before construction companies can fully utilize robots on their working sites. At present, leading contractors have been actively engaged in the development of robots designed to free human workers from unpleasant work, rather than to raise productivity. A major difference between construction robots and factory robots is the necessity of construction robots to move from one spot to another, for example, from the base of a structure to the top. Consequently, the assembly and fixing work of the construction site must be systematized to allow greater robotization.

* In Japan, settling disputes with residents is an important part of construction projects for clients and contractors. The causes of hard feelings among residents may be divided into the physical, social, and construction factors. The physical factor includes a loss of sunlight, increased air currents, and a disturbance of TV broadcast reception. The social factor includes possible changes in the flow of pedestrians and motor traffic (e.g., construction of a supermarket building will lure shoppers away from existing shopping areas) and unfavorable social images of the new building (e.g., a plan to build a game machine center will provoke education-conscious parents). The construction factor includes construction noise and vibration.

Table 2.2 Construction Robots Used by Japanese Contractors

	Shimizu	Kajima	Taisei
Steel structure work	• SSR-3: Spray robot for fireproofing • Mighty Jack: Remote assembly manipulator for steel beams • Mighty Shackle Ace: Radio control autorelease clamp		
Concrete work	• Multi-purpose traveling vehicle for concrete slab cleaning & grinding • Concrete slab finishing robot	• Concrete floor finishing robot	
Painting	• SALIS: Spray robot for concrete silo surface lining • OSR-1: Mobile spray robot for exterior wall painting		• Ultra high wall spraying robot
Site renovation			
Inspection		• Tile exfoliation detector	• Exterior tile inspection robot
Cleaning			
Automatic power station		• Reinforcement bar distribution robot • Stud welding robot	

Takenaka	Ohbayashi	Others
	• Autoclamp	
• Concrete horizontal distributor • Concrete distribution crane • Concrete floor finishing	• Bracing crane	• Toda: Bracing robot (Work Master) • Fujita: Jacking robot system
		• Tokyu Const.: Rubble masonry robot
• Exterior wall inspector	• Clean room inspection robot	• JGC Construction: In-pipe horizontal runner • JGC Construction: In-pipe runner • Toda: Clean room inspection robot • Tokyo Gas: In-pipe inspection robot
		• Meidensha: Duct cleaning robot • Toshiba: Automated floor sweeper • Mitsubishi Electric: Automated window wiper
		• Hitachi Plant Engineering: Air filter auto changer

Another way of improving productivity is to use a greater percentage of lightweight, modular prefabricated parts and components. Along with robotization, these measures will simplify construction jobs and, therefore, will alleviate the problems of the shortage and aging of skilled workers. If these technologies are applied systematically, it will become possible to achieve on construction sites the comfortable working conditions achieved in many manufacturing plants, and thus to attract young talented workers from other industries or straight from school. Structure work especially requires a new concept of production systems to weed out its particularly acute labor shortage. New technology-intensive production systems will not only increase productivity, but also differentiate construction techniques among companies through patent right protection and eventually alter the employment practices and subcontractor relations of each contractor, widening the gap between successful and unsuccessful construction companies (see Figure 2.12).

There are similar challenges in the field of installation work, where the installation worker cost increased more sharply than the increase in total construction expense. The required technical level of installation work has gone up, as exemplified by high-tech clean rooms. As a result, contractors are obligated to step up research and development efforts for installation techniques, including the modularization of pipes and fittings and the innovation of new jointing methods. Further, since equipment and fixtures for clean rooms have shorter life cycles than those of building structures, demand will expand for replacement of equipment and fixtures exclusive of building work. It will then become important to develop easy-installation systems ro reduce the installation labor cost and life cycle cost as part of the effort to improve productivity.

Using New Materials

Among the world's most earthquake-plagued countries, Japan in 1981 enacted the New Anti-Earthquake Architectural Design Law, requiring the use of new anti-quake design computation methods based on recent quake data. Under this law, a permissible stress design method is applied to buildings no more than 31 meters high; for higher buildings, the law requires the architects also to investigate the predictable plasticity of the designed structure beyond its flexibility limit in the event of a severe earthquake. Further, the design of buildings exceeding 60 meters in height must receive the approval of the Construction Minister.

Since the enactment of this law, the use of concrete, reinforcing bars, and steel frames per square meter of construction floor space has increased by as much as 10% (see Figure 2.13). If new materials having high strength but a smaller cross-section are developed, the input of materials can be reduced and the free space within the building can be increased to bring about a major change in building systems. Although not a structural material, use of a carbon fiber reinforced curtain wall developed jointly by Kajima and Sumitomo Metal Industries in 1985 has

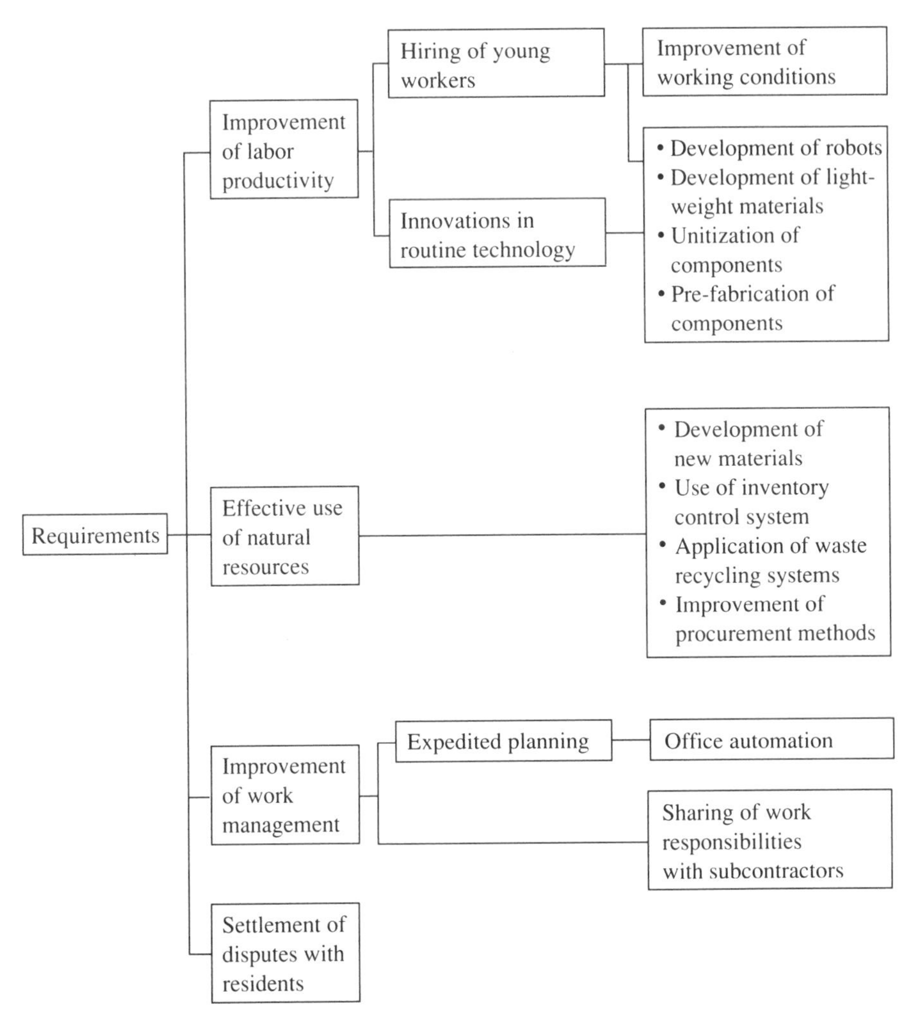

Figure 2.12 *Requirements for improvement of construction productivity.*

cut reinforcing bar requirements by 20%. A super-concrete structure developed by Shimizu Corporation to improve the strength and economic performance of super–high-rise buildings is now in the experimental stage; a high-strength concrete developed by Ohbayashi is also a promising new material.

In the interest of effective use of natural resources, the development of new materials can be better achieved if construction firms enter into technical collaboration with other industries and universities. An effective use of materials can also

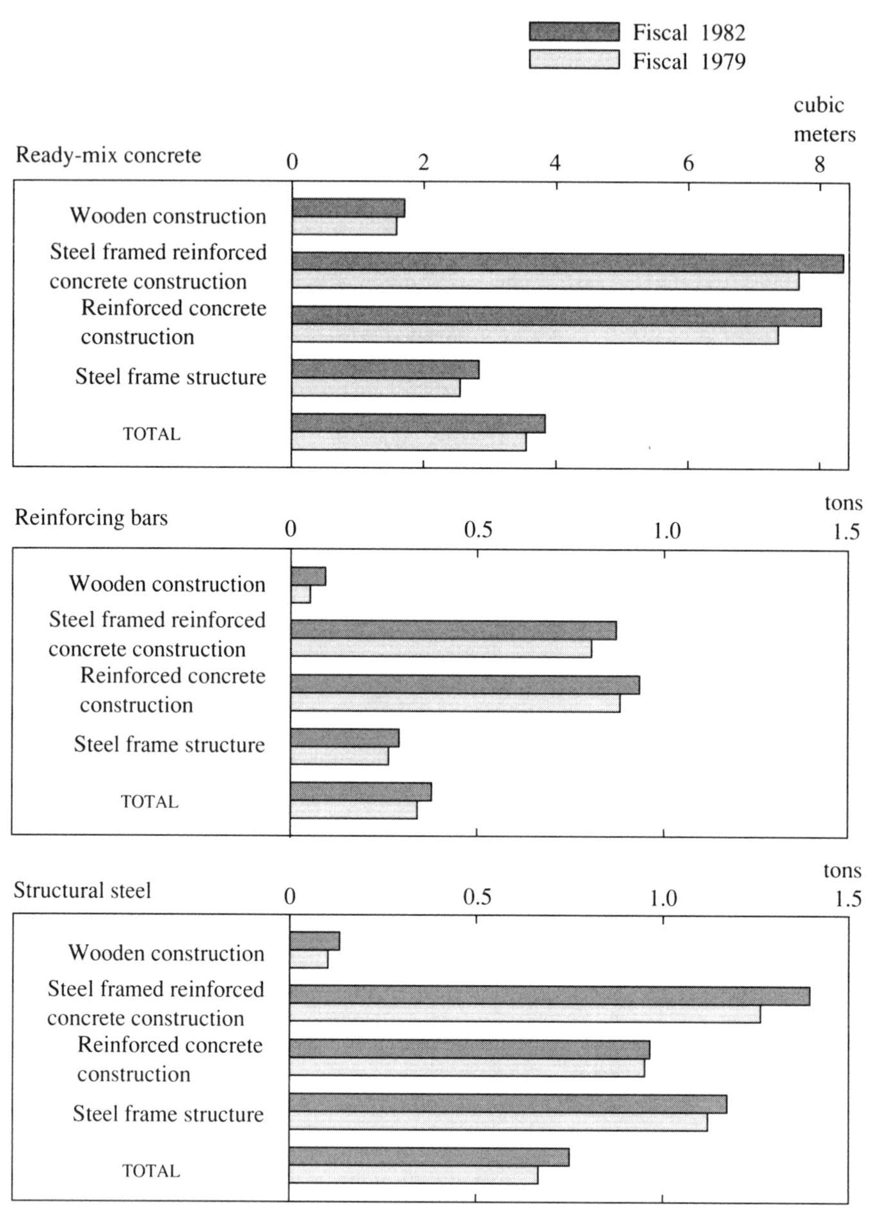

Source: Monthly Report on Construction Labor and Materials (Kensetsu Ro'odo'o Shizai Geppo'o), Ministry of Construction, January 1985.

Figure 2.13 Use of construction materials per 10 square meters

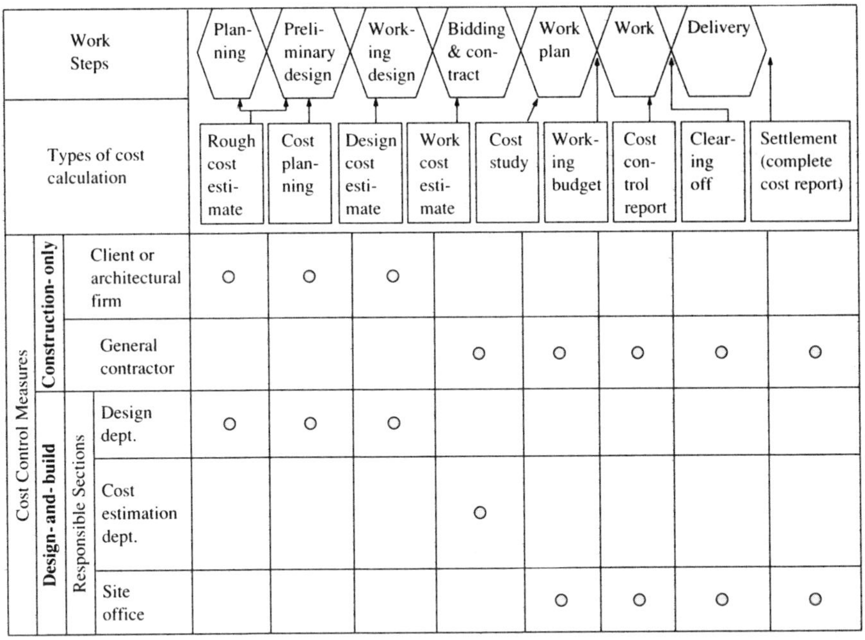

Work Steps			Planning	Preliminary design	Working design	Bidding & contract	Work plan	Work		Delivery
Types of cost calculation			Rough cost estimate	Cost planning	Design cost estimate	Work cost estimate / Cost study	Working budget	Cost control report	Clearing off	Settlement (complete cost report)
Cost Control Measures	Construction-only	Client or architectural firm	O	O	O					
		General contractor				O	O	O	O	O
	Design-and-build (Responsible Sections)	Design dept.	O	O	O					
		Cost estimation dept.				O				
		Site office					O	O	O	O

Figure 2.14 Cost control and work steps

be achieved by purchasing the materials directly from the producers; by improving inventory techniques through, for example, the introduction of a point-of-sale computer system; by developing methods of recycling waste materials; and by providing material producers with technical assistance in reducing material production costs. In short, there are many ways in which contractors can achieve effective use of construction material if they study an overall view of the material flow, from the maker to the dumping ground.

Elements of Work Management

The four most important elements in construction work management are quality, cost, delivery speed, and safety. The cost and schedule control methods prevalent in Japan are introduced below.

COST CONTROL. A different type of cost calculation is executed at each work step by the section concerned (Figure 2.14). In Japan, general contractors commonly sign contracts on a lump sum basis, and these contracts are divided into two types: construction-only and the design-and-build. In the first case, the planning, preliminary design, working design, and associated cost controls are carried out by the client or the commissioned architectural firm. For the general contractor, the work cost estimate produced by its cost estimation department for bidding purposes serves as the first data for cost control. Since the contract price is determined in these initial stages under the prevalent lump sum installment contract scheme, the work cost estimate is of vital importance to the general contractor.

Once the contract is signed, the cost control responsibility is transferred from the main office to the construction site office. This site office plans a working budget, allotting the necessary working funds for each type of work, and begins cost control activities on the basis of this budget plan. General contractors usually make each construction site office submit, throughout the construction period, monthly cost control reports recording specific payments and payable liabilities, so that the main office can check on overspending and can promptly take steps to correct excessive outlay.

In the case of design-and-build contracts, the general contractor's design department handles the planning, preliminary design, and working design according to the needs and budget of the client, and the prevailing contract method is again the lump sum arrangement. Once the contract takes effect, the site office becomes the responsible section for controlling costs.

SCHEDULE CONTROL. Control of the construction schedule and work procedures affects not only the delivery date but also the cost and quality of the construction work. The schedule control plan includes total, monthly, and 10-day schedule tables; it is produced by the site office manager and staff, using bar charts, network techniques, and other management tools. Because of the vital importance of cost and schedule control carried out by site offices, Japanese general contractors deploy top-level engineers at these field offices, and office managers are appointed from among leading personnel with 10 to 15 years of work experience in various construction sites. To facilitate the jobs of the office managers and staff, efforts are being made to increase the use of computers at the site offices.

Productivity in Work Management

General contractors can improve productivity by expediting the planning of an optimal work procedure and by upgrading work management. For the first objective, the contractor's accumulated know-how must be easily and rapidly available; this is best achieved by the efficient use of office automation equipment. Leading general contractors are developing their own office automation systems for the management of construction work. Ohbayashi, for one, has a *Site System* designed handle administrative work, including cost calculation, by computer. Takenaka Corporation uses a *Cosmos* computer system to control work offices in construction to sites, and Shimizu Corporation has introduced to the market a portable device, the *Electric Area Calculator*, capable of instantly computing areas and volumes at the construction site. Nevertheless, most computer systems used by contractors are intended to automate repetitive office work and stop short of providing simulations that would help to draw up the optimal construction work plans.

Work management can be upgraded by acquiring construction management capabilities. The construction management methods of the United States, however, are not necessarily suitable to the Japanese industry, for the following reasons: (1) It is not customary in Japan to pay a reward for specialized knowledge or know-how, and (2) since the client company has a staff of able workers, there is little need to hire independent construction managers. Still, to perform international, full turn-key, urban redevelopment, and other large-scale projects, Japanese contractors must hire some specialist construction managers. An upgrading of construction management capacity and office automation systems—plus development of construction robots, new materials, and lighter, modular, or prefabricated components—will bring great change in the structure of the construction industry.

Accordingly, the roles played by general and specialized construction companies will alter. For example, it may become possible to commission a specialized company to carry out all of the structural work, including concrete, forming, and reinforcement work; this will remove a good amount of management workload off the shoulders of contractors, and the remaining management work can be simplified by an improved office automation system. Pruning of management work, perhaps, is a prerequisite for Japanese contractors wishing to expand into more sophisticated management activities, such as mediation between interested parties in urban redevelopment projects, an expansion of overseas projects, and an advance into new fields of business.

Cultivating Specialized Construction Companies

A feature of the Japanese construction industry is that a large number of specialist companies act as subcontractors, supplying skilled and unskilled workers. For structural work alone, various subcontractors exist in the fields of forming, reinforcement, structural steel work, pile work, forging, earth work, and so on. The great majority of these specialist sub-contractors are small companies, and characteristic relationships between general and specialist construction firms may be summarized as follows:

- There are no large organizations, such as the labor unions in the United States, capable of supplying skilled labor in Japan.
- Multilayer business relationships exist in many areas in the Japanese construction industry, where the general contractor hires specialist subcontractors who in turn hire smaller specialist firms.

- Technical and work capabilities of subcontractors vary widely; structural work subcontractors are especially noted for their supply of unskilled labor and their lower than average management abilities.
- Specialist subcontractors with excellent work ability are much in demand by general contractors, most of whom are anxious to organize and bolster partnerships with able subcontractors.
- The strong interdependency of general contractors and specialist subcontractors has been responsible for much of the delay in the modernization of the Japanese construction industry.

To achieve efficient construction of high-quality buildings, specialist companies must modernize their operations. Structural work subcontractors, in particular, must become more independent, reliable builders and stop operating like manpower agents. General contractors hope that, instead of merely supplying the number of workers asked, these subcontractors will become equal, responsible partners able to take the initiative in managing assignments.

Tokyo-based Mukai Kensetsu Company is one major company specializing in piling and earth work which is now trying to evolve into a total structural work company. Toshio Mukai, president of the company, has listed the following items as important ingredients for future specialist construction firms:

1. Reform of top management's thinking
2. Strengthened organization for business operations
3. Use of statistical management techniques
4. Diversification of business

He also mentioned the following as vital ingredients for the improvement of construction work and management capabilities:

1. Quality assurance
2. Upgrading of equipment and machinery
3. Education of young technical workers
4. Ability to perform total structural work

Mr. Mukai made a particular point that the reform of top management's thinking is the most vital starting point for corporate evolution.

For specialist firms to become self-managing contractors, knowledge of the role played by general contractors can be extremely important. General contractors can provide much guidance to their subcontractors, in that they can raise the technical responsibilities of their subcontractors by specifying, for example, the finish standard and other criteria for satisfactory work completion. Also, general contractors can help subcontractors to work in an orderly manner by distributing orders as systematically as possible to minimize seasonal fluctuations. It would benefit both general contractors and subcontractors to work out a detailed cost breakdown of the subcontract work, including management fees. Particularly important, subcontractors must train their young technical workers, but many are so small that they cannot afford to offer an expensive education to their workers. General contractors are in the best position to offer both human and monetary assistance in education, such as holding workshops for the young employees of subcontractors.

MARKET-ORIENTED OPERATIONS

The M-Project

A project that is highly symbolic for the large general contractor is scheduled for completion in 1993. This is the M-Project, which involves 23 billion yen ($150 million) in private investments. The plan calls for the construction of a cluster of high-rise buildings housing a hotel, corporate showrooms, sports clubs, theaters, a television studio, fashion boutiques, arcades, and so on—a city within a city. This general contractor (name undisclosed) is the leading planner of the M-Project, which was conceived during regular study meetings held by the employees of the construction company and a number of banks. The meeting attendees agreed that large construction enterprises will no longer be profitable in big cities such as Tokyo if the necessary land must be purchased at an exorbitant price. The members, therefore, considered the utilization of *air space rights** as the most promising way to implement major construction projects in metropolitan areas. Thus, the M-Project became the first case in Japan where air space rights were applied to effect a construction project.

Upon securing an M-Project site, the general contractor acted as the planning coordinator, drawing up the construction plan to reflect the clients' desires and prospective tenants' business needs. As another feature of this project, the general contractor, together with clients, financial companies, and leading tenant corpora-

*The concept of air space rights, or *transferable development rights*, introduced to Japan from the United States, recognizes the air space above land as a marketable commodity and gives the owners of these air spaces the rights to build structures therein. The concept of air space rights has not yet been thoroughly fully established in Japan.

tions, formed a management company which will rent all of the M-Project premises from the client and will manage the buildings on a commercial basis. As a part owner of this company, the general contractor is committed not only as a building maintenance agent but also as a promoter and a marketer of these rental buildings, with the responsibility of attracting a maximum number of tenants and shoppers. Thus, unlike most construction work, the general contractor's job will not end when the M-Project construction work is completed.

Project-priming Business

The activities carried out by the above general contractor for the M-Project are summarized as follows:

- The contractor for the first time employed the air space rights technique in Japan to garner sufficient construction space. In this process, the contractor overcame a number of legal restrictions.
- The contractor took the leadership in the planning of the M-Project, starting from the formulation of fundamental business concepts to the specification of plan details.
- The contractor joined in the management of the M-Project premises, sharing with the owners the responsibility of finding tenants and bringing in business profits.
- As an owner of the management company, the contractor will stage various promotional events and launch advertising campaigns.

Until recently, all these activities would have been done by the building owners. The contractor has stretched the company's activities beyond the border of the traditional order-taking operation, and has assumed a new role as a project promoter or as a project producer. The construction work has become only the hardware part of the project, with the software part giving new business opportunities. The M-Project is a typical case of development business, and the prevailing view is that the percentage of these projects will increase, particularly large-scale projects.

Projects as Joint Enterprises

The word *project* may be defined as a successful undertaking requiring more than one individual or one organization (including a department of the company or an office of the government), or more than a conventional combination of individuals or organizations. In this regard, virtually all endeavors are potential projects. In the Japanese fashion world, for example, the Bigi Group has achieved phenomenal growth on the basis of the production of designer brand apparel. By contracting a number of established designers to market apparel products under its brand name, Bigi itself has become a brand to be reckoned with by fashion-conscious people. Similarly, the Comme des Garçons and Issey Miyake brands have enhanced their appeal by entering into partnership with store interior designers.

The increase of projects may be attributable to the increasingly complex makeup of the social fabric and the rise of individual life-styles in Japan. Things are changing faster now than before, and today even the shrewdest businesspeople have difficulty in forecasting the future. The business risk is correspondingly large, and projects have emerged as a means of spreading risks. In the case of construction work, the client must face a large risk in placing an order worth millions of dollars, and the risk appears larger today than before because of an increasing uncertainty about the future.

Take for instance a shopping center. When Japanese consumers were beginning to rise from postwar poverty to new affluence, it was comparatively easy for the client to determine the types of buildings and tenants needed to make the investment profitable. But recently, the client is not certain about the success of this formula, and prospective tenants, uncertain whether their rented stores will be profitable, are becoming reluctant to sign a tenant contract. As a result, the client often asks the contractor to find tenants or to cooperate in running the rental building business after the construction work is completed.

It may sound odd that clients who want to invest in new buildings are not always sure what type of building they hope to build, but this is a normal phenomenon in the business world. Executives do not know everything about their companies, so professional management consultants are hired. Office workers do not know exactly how their offices are run, so they rely on the computer maker's office automation software to improve office productivity. Government offices do not really know what projects are most effective for the improvement of the nation's social infrastructure and for the stimulation of domestic demand. The government therefore hopes to promote Minkatsu* construction projects intended to attract private funds and private ideas through financial and tax incentives. The plans to build a New Kansai International Airport off the shore of Osaka and a highway bridge across the Bay of Tokyo are two examples of Minkatsu projects with emphasis on private creativity.

* Minkatsu projects are those projects which, until recently, would have been carried out by the national or local governments but are now carried out by private companies, so that public infrastructures can be developed without increasing the government budgetary burdens. Minkatsu projects are considered an effective means of stimulating domestic demand. The planned construction of a new international airport off the shore of Osaka and a highway bridge across the Tokyo Bay are typical Minkatsu projects.

Management Resources of Contractors

Because things have become more uncertain and businesspeople have found it harder to read the future, projects (i.e., undertakings by more than one participant) are increasing and construction companies are called upon to develop projects from a blank sheet of paper. Do construction firms have more advantages than other industries in promoting development projects? If so, what are the specific management resources that are more advanced in construction companies than in other companies?

- First, construction companies possess technologies and construction work know-how. Although opportunities to move into other business areas are expanding, construction work will remain the main business of construction companies.
- Since contractors receive orders from all types of industries and government offices, they have an extensive network of information collected through business encounters. Only banks and general trading houses command equally extensive information networks. The contractors' high-level information-gathering ability is the key to leadership in construction projects.
- Through vast experience in mediation between clients and surrounding residents, contractors have acquired excellent mediation ability and insight, and this ability is essential to the management of a project participated in by several interested parties. Nevertheless, much of the mediation ability belongs to the individual employees assigned to mediation duty, and contractors must process these individual abilities into systematic company know-how which will be used for project promotion on a companywide, rather than individual basis.

Marketing Operations of Construction Companies

A comparison is made between development and traditional marketing activities in Table 2.3. The latter is the prevailing marketing style practiced throughout the postwar period up to the first oil crisis in late 1973. During this period, the clients' most important criterion for selecting a contractor was an ability to build high-quality buildings—the hardware. Clients found it easy to attract tenants for their office buildings and condominium apartments and to quickly recoup their investments in new factories and shopping centers. Therefore, given the basic ability to supply a building of reasonable quality at a reasonable price within a reasonable delivery time, the client selected a construction firm on the basis of long business relationships, social reputation, and the personalities of individual businessmen from the construction company.

Table 2.3 Comparison between development and traditional marketing

	Traditional Marketing	Development Marketing
Period	High economic growth period	Low economic growth period
Market environment	Easy to sell	Uncertain about sales outcome
Market needs	Basic hardware functions (quality, delivery time, etc.)	Commitment to project
Marketing tactics	• Solving of construction problems • Use of personal contacts • Friendship through entertainment • Emphasis on person-to-person confidence	(In addition to items on left) • Information service on real estate merchandise • Guarantee of tenant occupancy • Formulation of project concepts and participation in project • Management of project • Sharing of financial risks
Competition	Relatively moderate	Very tough

In response, contractors concentrated their attention on satisfying the clients' building needs, particularly the budgetary plans. Emphasis was also placed on relations with surrounding residents and local administrative offices in charge of construction regulations. When competing with rival contractors, the construction company was eager to expand friendly contacts in the business community, and to collect information about prospective clients before rival companies could get to them. Then, sales people worked hard to consolidate friendships with the prospective clients, inviting them to a golf course or to a tea house and sending expensive mid-summer and year-end gifts. They visited the would-be clients almost daily to win their confidence—a Japanese-style of sales approach based on personal relations. Even today, sales departments in Japanese construction companies are organized on a client-by-client basis. As a result, contractors seldom considered competition in terms of differentiating their products from those of competitors during the period of high economic growth.

After the first oil crisis, the market environment radically changed. The hitherto brisk growth of household incomes sharply decelerated, and consumers were compelled to select purchases with increased care, and to refrain from impulsive buying. The consumers' criteria for purchase selection came to be based on individual preferences. This forced corporations to supply a greater variety of products and services in smaller quantities, and to practice more detailed marketing activities. Corporations tried to meet the growing need for faster, more flexible marketing and product planning activities by using computers and communication equipment—which started the *Information Era*.

The construction market has changed accordingly. Discovering that tenants were no longer there for the taking, clients began to think of ways to differentiate their condominium buildings and shopping centers from those of competitors. Clients were anxious to select the best locations to ensure the success of their building investments. The change in client mentality was felt strongly by contractors. The builders were asked to find better located sites and guarantee a sufficient tenant turnout; in short, the contractors were obligated to share some of the clients' risks, in what the clients considered a spirit of give and take. In these situations, the long-standing client relations built by leading contractors were sometimes toppled by the bold risk-taking tactics of smaller contractors.

Faced with bleak market conditions, large general contractors realized that they lacked strategies to counter the arising difficulties. They noticed that their hardware technologies were more or less equal, and any leading contractor was able to construct super–high-rise buildings. They were alike, and there were few technologies or marketing techniques that would differentiate one from the other. Thus, the contractors became aware that the needs of the construction market were no longer only for structures (the hardware), but also for the contractor's commitment to clients' business, the key point being an ability to offer services that assist the business of the client (see Figure 2.15).

Possible Marketing Strategies

This will again point to the need for projects, or the need for combining a number of interested parties to provide client services and spread risks. One approach would be to formulate a marketing strategy technology. To build commodity distribution facilities, for example, it would be best for contractors to collaborate with a manufacturer of material handling equipment. Partnership with a manufacturer of computers and communication equipment would be a big help when trying to win orders for intelligent buildings. Because of its freedom in selecting technical partners on a project-by-project basis, the construction company may be the ultimate builder of a computer-intensive flexible manufacturing system for reducing human labor in factory operations.

The second marketing strategy is to emphasize architectural designs. Construction has always reflected the cultural aspirations of each age, and it would be a rich contribution to build architectural structures that express today's cultural needs. The American post-modernist architecture is a retreat from the pragmatism of modern architecture, incorporating historical legacies as design motifs and pro-

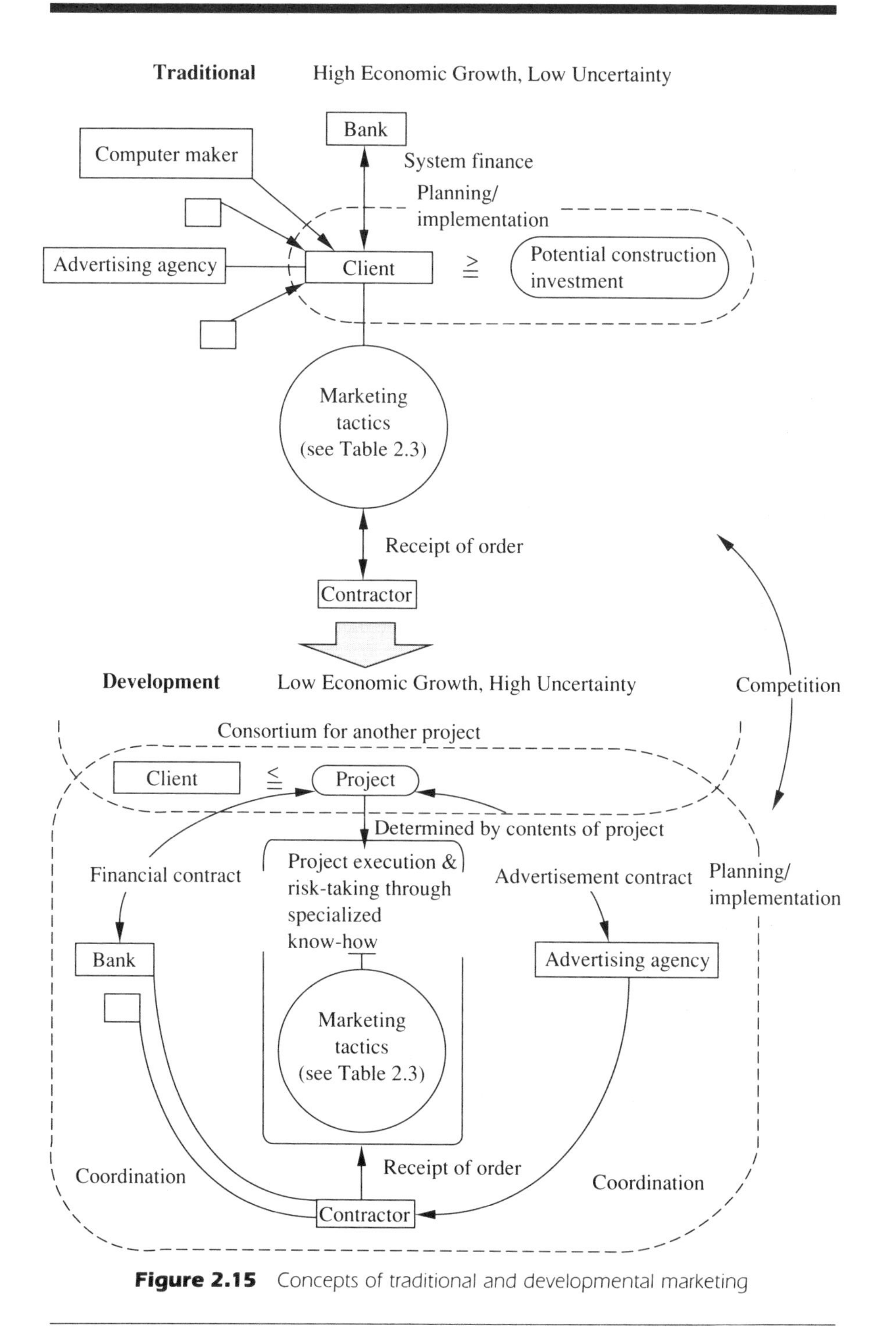

Figure 2.15 Concepts of traditional and developmental marketing

viding extra space and decoration at the expense of function. Post-modernism in Japanese commercial and condominium buildings would be a promising approach to the differentiation of architectural products. For clients, it would be advantageous in attracting shoppers and tenants to uniquely designed buildings, provided that design differentiation did not demand a large extra cost on top of the already exorbitant land price. Further, the contractor will have to increase public relations and cultural activities to inform the general public of the contractor's design capabilities.

In a third marketing strategy, the contractor may offer know-how to the client that will be helpful to the client's business. For the owners of commercial buildings, knowledge of customer marketing techniques will be welcome. For clients who want to build hotels, expertise on hotel services may be offered. These projects are similar to the aforementioned M-Project in customer services.

A fourth market strategy is to shoulder a maximum amount of financial risk: guarantee a total tenant occupancy, for example, or purchase some condominiums from the client who has ordered a condominium building. Considering the financial fluidity of the contracting business, contractors should perhaps avoid the easy policy of shouldering monetary risks for clients.

There should be other viable strategies, or mixtures of different strategies. Yet all strategies will have construction as the core technology and will be aimed at the same goals:

1. Expansion of the business domain
2. Increase of the added value

For the first goal, construction companies must compete with other industries if their business activities expand outward from the construction industry proper. If a contractor is to undertake a shopping complex project, the company may team up with a distribution company and an advertising agency. However, it should be kept in mind that, although they are partners in this particular project, they may become competitors in other projects. In this connection, the Seibu Saison Group, a distribution conglomerate with total annual sales of about 20 billion dollars, reportedly plans to start an urban redevelopment business, taking advantage of its retailing know-how. To the general contractor, Seibu appears more foe than friend.

Similarly, in contractors are to advance into the designing of flexible manufacturing systems for plant operators, they will eventually compete with computer and machinery makers. Takasago Thermal Engineering Company is a specific machinery maker which is already in competition with construction companies in the area of high-level clean room facilities. The technological clout of NTT (Nippon Telegraph and Telephone), the world's largest communications company, has already spread beyond the borders of electronic communication to the outskirts of the construction market. NTT is the corporation with the largest number of employees with first-class construction engineer licenses and is now beginning to sell design and construction services for intelligent buildings.

To make matters more complicated, these potential competitors are, at this moment, also clients, and if the business domain of the contractors expands further, they will be obligated to abandon part of their market. Obviously, they will find it most advantageous to differentiate their new business from the business of potential competitors in different industries to minimize collisions. It would be more appropriate for the general contractors to aim at specific market areas than to diversify in unspecific general directions in new business fields.

The Construction Market in the Future

The first development construction projects suffered under the economic slump caused by the first oil shock. The underlying thought was that construction companies must use their own initiative to create demand. At present, although there are still very few true development projects, the number of such projects will no doubt increase in view of the predicted advance of the Information Era and consumer refinement of individual tastes and preferences (Figure 2.16). The larger the construction contract, the greater the development factor in the project. This

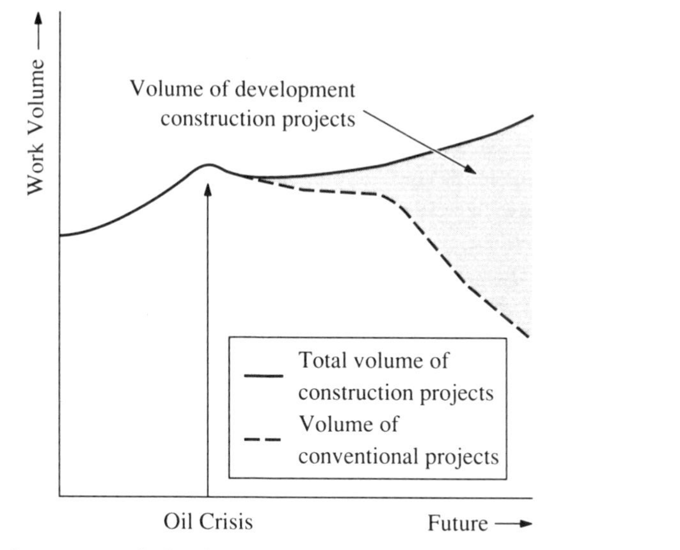

Figure 2.16 Percentage of development construction projects as envisioned by a general contractor

will bring more advantages than disadvantages to construction companies, whose business is usually over when the building is completed (except when there are complaints from former clients).

Since new orders are hard to find these days, however, it would be beneficial to retain a business relationship with the client beyond the completion of construction, which will increase the chances for winning maintenance jobs and remodeling work in the future. Further, considering the fact that the building is not the end object of the client, but is a means of carrying out a business operation, the contractor should find out the client's true business needs through after-sale relations, to enable him to offer value-added services to the client.

Long-Term Warranty on Buildings

When after-sale services are gradually upgraded, interest in quality guarantees on buildings also grows. Japanese cars enjoy a worldwide reputation for their high quality, and bear a quality warranty certificate issued by makers. Consumers take it for granted that automobiles, electronic home appliances, and other durables are protected by a warranty. On the contrary, it is questionable whether consumers are satisfied with the warranty on their new homes, the biggest purchase of their life. Condominiums of reinforced concrete generally provide a two-year warranty on damage, and in addition, there are often standard after-service provisions attached to the home purchasing contract, giving as a rule a five-year warranty on rain leakage and a three-year guarantee on defects in structural components. Comparing the six-year legal life for cars and a legal life of 60 or 65 years for concrete buildings, purchasers of this most expensive merchandise of a lifetime feel that these warranty periods are too short.

With this as the background, some contractors have already introduced longer warranties. Among single-unit home builders using traditional Japanese home-building techniques, those builders who have registered at the Registration Organization for Warrantied Houses, an office administered by the Construction Ministry, provide a standard long-term warranty: a 10-year guarantee on the durability of structural components, including foundations, floors, walls, and roofs, and a five-year warranty on the antileakage performance of roofs. Prefabricated home builders, who compete with the above home builders, have also provided a 10-year protection on structural components over the past several years. Some condominium builders have started a similar 10-year guarantee. In response, the Housing and Urban Development Corporation, the government-managed home supplier, has begun a long-term warranty program for some condominiums since 1983, as shown in Table 2.4.

Before this long-term warranty of houses and buildings can spread to all builders, a number of problems must be solved regarding design responsibilities, insurance systems, and business profitability among others. The fundamental idea behind long-term warranty is not to "guarantee free repair services for 10 years," but to "produce structures in which defects will not occur for at least 10 years." Since this puts greater importance on quality, a long-term warranty is a necessity for all construction companies and an inevitable outcome in today's quality-conscious market. Those companies unable to offer such a warranty will eventually lose out in the market competition.

Personnel for Project-creating Business

To implement project-creating marketing strategies, a far wider variety of business specialists will be needed. Commercial projects require marketing experts, and projects for advanced automated factories and intelligent buildings demand talents

Table 2.4 Long-term Housing Warranty by Japanese Housing and Urban Development Corporation

Area Covered	Applicable Damage	Warranty period
Building structure (including foundation, roof, balcony, stairs, and eaves)	Distortion or damage affecting strength of building structure	10
Roof	Rain leakage, and damage to building due to rain leakage	10
Exterior wall (including areas of contact with doors and windows)	Same as above	7
Water tank, high-placement water tank, sewage cleansing tank (including tank supports)	Rain leakage, and damage to building (including tanks) due to rain leakage	5
	Distortion or damage affecting strength of building structure	10
Bathroom	Rain leakage, and damage to building due to rain leakage	5
Antitermite treatment area of 1st-story wooden floor	Appearance of termites, and damage caused to building by termites (excluding cases where appearance of termites is known but not reported)	10

Source: Kazuo Shimotaka, "Long-term Guarantee on Quality by Japanese Housing and Development Corporation," *Construction Work (Seko'o)*, Sho'okokusha, July 1985.

in system engineering and data processing. All these people, nevertheless, must possess an excellent project planning capability, since they will be frequently meeting with the planning and financial staffs of other companies, and their creative ideas and actions will give them the lead in the promotion of projects. In addition, they must be knowledgeable in the affairs of their partners' industries. Further, it would be advantageous for them to have personal contacts with many types of people outside the construction industry.

The businessman who was assigned to the M-Project by the construction company responsible had an impressive knowledge of financial matters and was able to win the confidence of the people from banks participating in the project. When the members gave the green light on the project, the businessman presented to the client a plan concept for the M-Project and proposed specific promotional activities aimed at the general public. It was evident to all the attendees at the presentation meeting that he closely followed and had an analytical insight into fashion trends, shopping behavior, and other recent social developments; thanks to his expert performance, his construction company was able to take the leadership in the M-Project.

Organization for Development Business

Even though talented staff are secured, their abilities must be applied in such a way that the capability of the company will be expanded. To realize the maximum cumulative effect of individual employees, the company needs to reorganize its marketing organization, which has centered around clients to collect information about possible new orders faster than the competitors. Due to the increased complexity of construction projects, it has become difficult for each businessman to cover a number of different clients and projects simultaneously, and the new organization would be formed around different types of services to be offered to clients in development projects, rather than around different clients or groups of clients.

As a result, it would be better to divide the marketing organization into a conventional client-based unit and a new project promotion unit. The businesspeople in the first unit will continue to build close relations with clients and act as mediators between the client and the company's design and construction sections. The businesspeople in the second unit will be assigned to plan and promote specific projects, fully utilizing their specialized knowledge. Although these specialists have always existed, they have rarely appeared on the marketing front, but have merely provided necessary support whenever needed by the salesmen. Yet in the future development marketing organization, they should play a principal role, along with the sales staff, as the leading promoters of construction projects. In this regard, a number of large general contractors have already set up project promotion sections within their marketing departments.

CULTIVATING HUMAN RESOURCES

Workers with Different Talents

The types of employees sought by construction companies are rapidly changing: applied chemistry, artificial intelligence, mathematical principles, agriculture, biology, international relations, and marketing are some of the majors of university graduates employed in 1987. Employing graduates with some of these majors would have been unthinkable just a few years ago, but today construction companies are particularly keen to recruit these diverse talents. In Shimizu Corporation, for instance, nearly 70% of new employees hired straight from the university 10 years ago had majored in construction and civil engineering. Today, that has shrunk to 50% with a corresponding increase in electrical and mechanical engineering as well as arts majors. Even among the construction and engineering majors, the percentage of people from graduate school has increased at a remarkable pace. Other leading contractors have also adopted employment policies similar to that of Shimizu Corporation.

For the construction company to forge ahead with future strategies, strong management leadership is essential, and the recruitment and training of employees who will work on the front line in the market is also of vital importance. To develop strong, attractive merchandise in electronics- and biotechnology-related facilities and information-oriented offices, housing, and stores, engineers with high-tech knowledge are a necessity. Furthermore, the employment of liberal arts students is now growing, as contractors are anxious to branch into international operations, development projects, and a variety of nonconstruction businesses. As a consequence, in addition to law and economics majors, an increasing number of young people with majors in marketing, international relations, and the like are joining the construction companies.

New Personalities in Demand

In addition to diverse educational fields, construction companies are also looking for people with specialized knowledge and individual personalities. Construction was once basically a passive contract business involving minimum risk on the part of the contractor and minimum motivation to develop specialized merchandise. Construction companies have always valued workers with site work experience, a steady working performance, and a personality fitting in with the company hierarchy. Today, however, contractors are seeking different types of people—people with solid individual personalities. Personnel department directors of leading general contractors state unanimously that they are looking for persons with "an international perspective," "an entrepreneurial spirit," and "an active, aggressive personality."

What jobs will these people do in the future? Certainly, overseas assignments will increase, requiring them to live and represent their company in a foreign country for 10 years or even more. They will be expected to handle most jobs without direction from the main company, and sometimes will be forced to grapple with difficulties requiring a knowledge of local customs and practices. In the area of new business, workers will be expected to accurately grasp the changing needs of the market and society and may even organize a new business, overcoming any opposition from their cautious superiors in the company. In the construction market, future workers will not passively wait for orders to come in but will draw up and propose construction plans to prospective clients, acting also as organizers of joint enterprises with companies in other industries.

Recruitment and Revitalization of Middle-Age Workers

Japanese construction companies are steadily recruiting new types of employees, but those employees are still young and it will be years before they can be given important assignments. Contractors must acquire ready-to-use talent as soon as possible. To this end, the recruitment of personnel from different industries, such as new materials, electronics, biotechnology, trading, and realty businesses, is noticeably increasing, and contractors are also increasing the employment of foreign workers both abroad and within Japan. Overseas, the employment of engineers and accounting and contract clerks, in addition to construction workers, is growing. In Japan, though still small in number, foreign nationals are hired to support the company's international operations and to assume a key role in the marketing, planning, and design of factories and offices for foreign capital companies operating in Japan.

Above all, construction companies regard a revitalization of their middle-aged workers as the most important step to quickly replenish their work force with ready-to-use talent. Reeducation of employees in their thirties is most crucial, since they are still young enough to absorb new specialist knowledge, international thinking, an entrepreneurial spirit, and project organization and management know-how. Their age ensures that they will be the most active and responsible group of employees at the turn of the twenty-first century.

Mid-career training and schooling programs will be expanded. In-house training, which at present concentrates on obtaining similar skills and harmonious personalities, will now emphasize the mastering of diverse specialist skills on an individual basis. Opportunities to attend outside-company training classes will increase, and education programs will become options, rather than requirements. The dispatch of employees to graduate schools and other similar institutions in Japan and abroad is increasing. This, too, will become more of a voluntary endeavor by each employee; at Shimizu Corporation, any aspiring employee who has passed qualification exams will be assisted by the company to study at, for example, the Harvard Business School or MIT. As a practical training method, Japanese general contractors dispatch employees to banks, manufacturing companies, research institutes, and government offices for two or three years. This has proved helpful in gaining nonconstruction knowledge and building personal relations with other industries. The education of middle-aged workers, plus the recruiting of outside talent, will be a vital step in a successful escape from the Ice Age.

PART TWO
PASSPORT TO GROWTH — SIX STRATEGIES

WHY SIX STRATEGIES?

In the Japanese business community, there is a time-tested saying: "A period of change is a period of chance." Today, all Japanese companies are straining to learn about the changes that are taking place in society and the business world outside of their own industries, because changes mean new business opportunities. (Construction companies, of course, are never unaware of outside developments.) The gradually unveiling reality is not totally joyful. It now appears that the new market expected from technological innovations will not be so big as was thought. Rather, because so many future-minded companies have moved in, most of these new markets have already proved overly prone to price competition, allowing only small profits to competing companies.

The market for super-LSI chips is a classic example—a new market that was quickly glutted and plagued by a cutthroat underpricing competition. Another example is the new communications market, including CAPTAIN (Character and Pattern Telephone Access Information Network) and the cable television systems, both of which were considered until recently to symbolize the beginning of a high-level information era. The reality has shown that few homes are interested in these new information and broadcasting services because of the high costs and the poor quality of information and programs. Also, new materials have proven to be underdeveloped in application because they can be used only as substitutes for existing materials. If the present situation continues, the new materials market will most likely shrink from its current scale. Similarly, the market for biotechnological products has shown an unexpectedly modest growth. To add to this dismal outlook, foreign countries are now exerting pressure for a restriction of the imports of Japanese high-tech products.

Flanked by rapid technological progress on one hand and the sluggish responses of the actual market on the other, companies must attempt to continue stable growth while revising their management methods to keep pace with the technological advances. That is, on one hand, corporations must raise the efficiency of their existing business operations to the highest possible degree and, on the other hand, must expand into peripheral or totally new business fields generated by new technologies. The success or failure of corporations will depend not only on how well they can learn from their experiences but also on how well they challenge the unknown, the unexperienced. Past successes may not be repeatable in the future, and corporations are compelled to skillfully utilize and combine two opposing approaches—competition and cooperation with other companies.

Despite the disappointingly small markets created by new technologies so far, the majority of Japanese companies, including construction firms, believe in the importance of acquiring advanced technologies to achieve future growth. But construction companies, perhaps, harbor a more sober sense of the future, because the prevalent prediction says that the long-term growth of the construction market in Japan will not be strong. To deal with today's changing business environment and the uncertain future, the six different strategies introduced in the following six chapters may be utilized. If the traditional style of being a "contractor"—a made-to-order craftsman—was the formula for growth in the past, these six strategies may be considered a formula for future growth, a gateway to becoming a demand creator.

High-rise Residential Complex in Lot No. 10 (Iraq) Shimizu Corporation

CHAPTER 3

TRANSNATIONAL STRATEGY

WHY GO ABROAD?

Community Construction in Wartime

On August 9, 1979, a telex reached the main office of Shimizu Corporation in Tokyo bearing a message from Iraq, reporting the successful receipt of the "Iraq High-rise" order, a full turn-key $530 million regional development project, by the Japanese team of Mitsubishi Corporation (general trading house) and Shimizu Corporation. The Iraqi project proved to be a landmark for the transnational operations of Japanese construction companies during the first half of the 1980s. Located in downtown Baghdad on the west bank of the Tigris, the Iraq High-rise Project involved the construction of a complex of 2300 high-rise condominium, commercial, school, and multipurpose buildings, all fully air conditioned by a community heating and cooling system.

The construction work began in October 1979. Nearly a year later, in October 1980, the Iran-Iraq war broke out. But despite the war, the project was successfully completed in February 1984, making full use of Iraqi funds, Japanese technologies, and a labor force from India, Pakistan, Bangladesh, and the Philippines. The Radio Iraq overseas broadcasting service dubbed the Iraq High-rise as the Middle East's number one new town. The supermodern condominiums were sought by Baghdad citizens, and the auction sales were completed within three or four months after completion of the work. Today, the Iraqi people reportedly esteem the Iraqi High-rise as a social development project that has contributed greatly to the improvement of the infrastructure of their capital city.

History of Overseas Operations

Japanese contractors began to carry out overseas construction work during the postwar years as part of the Japanese national war reparation and economic cooperation programs. Only a very small number of projects were obtained on a commercial basis, and, except for these government-led projects, Japanese construction remained a typically domestic industry. A full-scale advance into the international market began only after the first oil crisis in 1973. At that time, Japanese contractors were literally forced to look outward by a sharp plunge in domestic orders, shrinking profits, a slump in home building, and a slash in government public works budgets, all in the wake of the oil embargo and the oil price spiral.

Soon, the contractors' outbound advance was helped by outgoing Japanese manufacturing companies. Car, electronic appliance, and industrial machinery makers began to expand their manufacturing operations abroad, mainly to take advantage of cheaper labor. Plant makers and petrochemical, chemical, steel, aluminum, and other intermediate industrial material producers moved some of their capacities to other countries to obtain cheaper fuels and materials. Service companies also opened department stores, hotels, and the like to cushion the domestic slump. In addition, there was increased demand for construction work in the Middle East, where vast amounts of oil money were available. Also, newly industrializing countries (NICs) in Asia developed an increasing need to improve their social infrastructure in step with their economic development.

As a result, the orders received from abroad chalked up a phenomenal growth of from 73 billion yen ($0.49 billion) in fiscal 1973 to 330 billion yen ($2.2 billion) in fiscal 1975, 630 billion yen ($4.2 billion) in fiscal 1979, and over one trillion yen ($6.7 billion) in fiscal 1983, as shown in Figure 3.1. Overseas orders have since hovered at around the one trillion yen level. A major share of the orders always came from Asian countries (Figure 3.2). Orders from the Middle East dropped from a 56% share in fiscal 1979 to as low as 5.4% because of an oil glut, slackening oil prices, the unending Iran-Iraq War, and a curb on developmental investments. In sharp contrast, orders from North America and Oceania have surged in recent years, mainly because of an increasing number of factories needed by Japanese manufacturers in these regions.

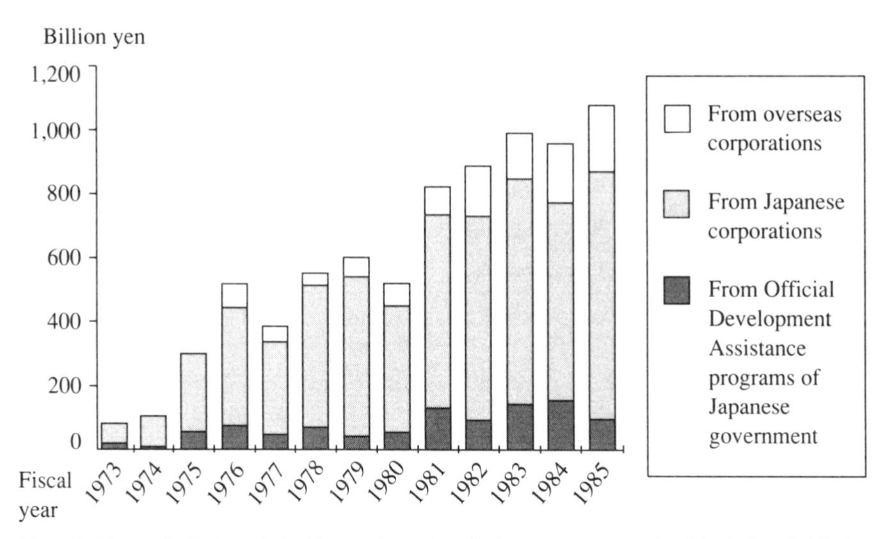

Note: 1. Figures include technical instruction prices. 2. "overseas corporations" include subsidiaries of Japanese corporations. 3. ODA programs include projects ordered by international organizations using funds from the Japanese government.

Source: Survey Report on Orders Received from Overseas Construction Work (Kaigai Kensetsuko'oji Juchu'u Cho'osakekka) (Tokyo: Ministry of Construction, 1986).

Figure 3.1 Value of overseas orders received by Japanese builders

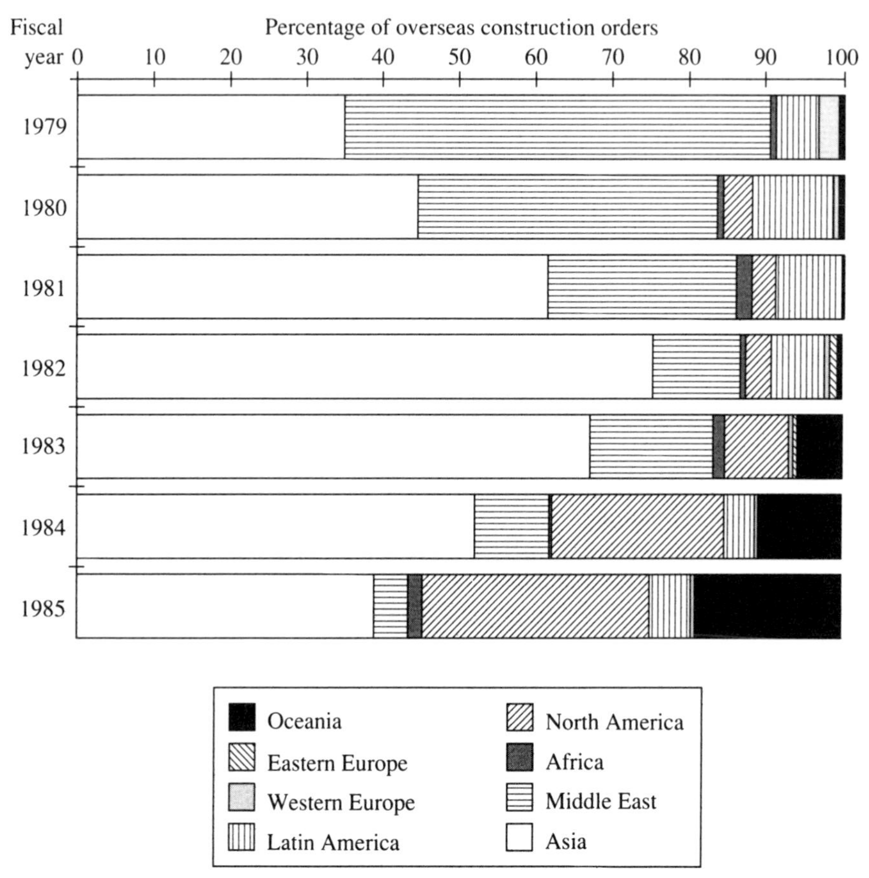

Source: Results of a survey by the Overseas Construction Association of Japan, Inc. on its member companies.

Figure 3.2 Overseas construction orders by region

Japan's Standing in the World Construction Industry

Although Japanese contractors have multiplied the number of orders received from abroad since the early 1970s, the scale of their international operations is still small compared with those of contractors in the United States, Europe, and South Korea. According to the latest data on the world's leading contractors, in terms of total domestic and overseas contracts, six Japanese contractors appeared among the top 12, led by seventh-ranked Kajima, eighth-ranked Shimizu Corporation, and ninth-ranked Kumagai Gumi in 1985 (Table 3.1). But when only contracts with overseas clients were compared, Japanese contractors were ranked much lower: Kumagai Gumi exceptionally ranked sixth, but Kajima 30th, Shimizu Corporation 34th, and so on down the list (Table 3.2).

For most leading contractors in Japan, foreign orders never account for more than 20% of their total orders, with some exceptions where the share reached about 30 to 40% of total orders. Foreign contractors generally gain a greater percentage of orders from overseas than do Japanese contractors. Take the world's largest contractor, Parsons of the United States, which garnered 58% of its orders from abroad; the second-ranked Bechtel Group posted 49%, and the third-ranked, M. W. Kellogg, scooped a whopping 92% as of 1985. Japanese contractors are deeply embedded in their comfortable 330 billion dollar domestic market, but, exactly because of this factor, they do have a large dormant potential to expand into the world market. Not surprisingly, some Japanese contractors may garner more orders from abroad than from inside Japan in the not-too-distant future.

Table 3.1 The Top 20 International Contractors by Total Contracts in 1985 (Millions of dollars)

Rank	Firm	Nationality	Total Contracts
1	The Parsons' Corp.	US	8,620.0
2	Bechtel Group	US	7,364.0
3	The M.W. Kellogg Co.	US	6,757.0
4	Morrison-Knudsen Corp.	US	5,887.7
5	Brown & Roots	US	5,578.7
6	Fluor Corp.	US	5,127.4
7	Kajima Corp.	Japan	4,953.4
8	Shimizu Corp.	Japan	4,779.1
9	Kumagai Gumi Co.	Japan	4,692.2
10	Taisei Corp.	Japan	4,641.3
11	Ohbayashi Corp.	Japan	4,280.0
12	Takenaka Corp.	Japan	3,787.0
13	Lummus Crest	US	3,500.0
14	Philipp Holzman	W. Germany	3,232.5
15	Stearns Catalytic Corp.	US	3,118.9
16	Bouygues	France	3,050.0
17	Hyundai Engrg. & Const. Co.	Korea	2,536.4
18	SAE-Socitété Auxiliare d'Entreprises	France	2,501.0
19	Spie Batignolles	France	2,442.6
20	Hazama Gumi	Japan	2,266.3

Source: "Top International Contractors," *Engineering News Records,* (McGraw-Hill, July 17, 1986) 38 +.

Table 3.2 Foreign Contracts of Top International Contractors in 1985

Rank	Firm	Nationality	Foreign Contracts (Millions of dollars)	Share in total
1	The M.W. Kellogg Co.	US	6,204.0	91.8%
2	The Parsons Corp.	US	5,027.1	58.3
3	Bechtel Group	US	3,626.0	49.2
4	Brown & Root	US	2,928.7	52.5
5	Lummus Crest	US	2,420.0	69.1
6	Kumogai Gumi Co.	Japan	2,175.8	46.4
7	Hyundai Engrg. & Const. Co.	Korea	1,998.1	78.8
8	Philipp Holzman	W. Germany	1,948.2	60.3
9	SADE/SADELMI Group of Cos.	Italy	1,795.4	99.8
10	John Brown Engrs & Contractors	England	1,743.0	84.4
11	Davy Corp.	England	1,611.3	76.0
12	Foster Wheeler Corp.	US	1,426.0	69.6
13	Fluor Corp	US	1,392.2	27.2
14	IMPRESIT SpA	Italy	1,345.9	82.4
15	Bouygues	France	1,250.0	41.0
16	Spie Batignolles	France	1,212.3	49.6
17	Italimpianti SpA	Italy	1,170.5	84.3
18	SAE-Société Auxiliare d'Entreprises	France	1,164.0	46.5
19	Enka Const. & Industry Co. Inc.	Turkey	1,081.6	68.8
20	JGC Corp.	Japan	1,020.1	71.8
30	Kajima Corp.	Japan	627.7	12.7
34	Shimiza Corp.	Japan	605.4	12.7
39	Hazama Gumi	Japan	499.8	22.1
50	Ohbayashi Corp.	Japan	377.8	8.8
65	Taisei Corp.	Japan	266.5	5.7

Source: "Top International Contractors," *Engineering News Records,* (McGraw-Hill, July 17, 1986) 38+.

Care in Construction Export

Export activities of construction companies are different from the exports of automobiles, electronic home appliances, precision machinery, steel products, and other commodities in that the former involve the transfer of knowledge, technologies, and know-how in a much purer form. Construction exports always result in local construction work, thus contributing to the social and economic development of the host country through the building of civil structures, cities, and industrial facilities. Construction exports also require a large percentage of the necessary labor and materials to be procured from the local market or from third-country markets, making it less likely to cause trade conflicts with host nations. Nevertheless, an unrestrained inroad into foreign markets could threaten local contractors, and it would be more appropriate to enter foreign markets in more locally ingrained ways, such as by setting up a local corporation, hiring local staff, and sharing technologies and know-how with local construction firms.

HOW TO BECOME TRANSNATIONAL SYSTEM ORGANIZERS

Contractors as System Organizers

The trading, banking, auto, electronic machinery, and industrial plant companies are generally considered most able to become worldwide system organizers* in the twenty-first century. Yet construction companies also have sufficient skill to organize international projects, although their overseas experiences are still very limited. The aforementioned Iraq High-rise Project, for example, combined a $500 million fund from the Iraqi government with Japanese technologies and project management capabilities from Japan, and labor and material supplies from Iraq and a number of other countries. Thus, the Japanese contractor (Shimizu Corporation) acted as a "redistributor of wealth," paying various countries for the labor and materials supplied—which is an important task of a system organizer. Several Japanese contractors, besides Shimizu, have successfully assumed an international system organizer role.

In order for Japanese contractors to act as system organizers in a greater number of international projects in the future, the following abilities must be acquired or improved:

- Collection and analysis of information on the social, religious, cultural, economic, and political environments of the foreign country concerned

*System Organizer: *System* is defined as a set or arrangement of things so connected and operated as to achieve an initial goal. The system organizer in a construction project coordinates different companies from different industries to perform various assignments, including the development of market needs, planning, design, construction work, procurement of machinery and materials, and the maintenance and administration of the buildings, in such a way that the objective of a construction project can be achieved. Japanese contractors hope to serve as system organizers in overseas projects.

- Analysis and coordination of a project on the basis of an insight into human behavior and the local philosophy of life
- Consulting
- Design
- Research and planning
- Procurement of labor, funds, and materials from different countries
- Project management
- Operation and maintenance
- Research and development related to overseas projects

Although Japanese contractors have proven to be efficient system organizers in the domestic market, they will have to polish up the above capabilities to become qualified system organizers in the international market. This can be done only by revising the prevalent concept that overseas operations are the specialty of the company's international department, and by realigning the entire company organization to transnational activity. In this regard, the whole Japanese industrial structure has incorporated international elements largely as a result of expanded export activities, and this has influenced many individual corporations to acquire overseas operation capabilities. Construction companies are finally responding to this call.

The Harsh Business Environment Abroad

In their attempts to become experienced international system organizers, what are the possible obstacles lying in wait for Japanese contractors? One type of stumbling block may be shown by a Southeast Asian example. Some time ago, the market demand declined in Southeast Asia, and Japanese contractors began to take orders for small-scale projects that had been the exclusive domain of small local construction firms. This caused flurries of protests from local governments and contractors. To prevent the occurrence of this type of problem, or to foster local industries and reduce the outflow of foreign currencies, the governments of developing countries usually limit the type and scope of orders, inviting foreign contractors to tender bids only for projects that are too technically or financially demanding for local contractors.

The Saudi Arabian government, for one, obligates all foreign contractors to pass on to local subcontractors at least 30% of the amount of an order obtained in Saudi Arabia. Malaysia forges ahead with its Bumiputra policy, intended to upgrade the socioeconomic conditions of native Malaysians. It not only puts a curb on foreign capital participation and employment of foreign workers, but also limits the scope of construction work permitted to non-Malaysian contractors. In Singapore, Hong Kong, and elsewhere, foreign contractors are obligated to accept the appointment of specific local subcontractors at the time of signing the contract, and the foreign contractor is given only minimum attendance fees.*

Some countries require the use of locally produced machinery and materials or low-priced machinery and materials available from third countries, which are supplied by the client after the contract is signed. This eliminates the profit that the contractor would normally earn from procurement activities. As one of the countries raising nontariff barriers to the activities of foreign contractors, Kuwait, in principle, prohibits the importation of pipes and industrial machinery. As a result, foreign contractors have had difficulty in bringing in pipes for temporary facilities as well as high-speed cutters, tapping machines, and other machine tools from their home countries. Further, some countries insist that the contractor accept payment in oil, or import or sell a certain amount of local products in exchange for the contract. These restrictions are aimed at giving maximum business opportunities to local companies and promoting joint ventures between foreign and local companies to stimulate a transfer of technologies from haves to have-nots. Overall, the restrictions against foreign construction companies are growing, and Japanese contractors must overcome these obstacles to evolve into capable system organizers able to operate on a worldwide basis.

Three Files at the Al Basrah Customhouse

One Japanese construction employee stationed in Iraq in particular has experienced the difficulty of dealing with local laws and regulations. On April 3, 1984, this Japanese businessman, handling the customs affairs of a Japanese construction company, was summoned by the court section of the Al Basrah customhouse. Accompanied by a local lawyer, the Japanese took a seven-hour drive from Baghdad to the port town situated at the junction of the Tigris and the Euphrates Rivers near the Persian Gulf. The Iran-Iraq War had broken out two months earlier, and, as it is only about 19 miles from the war front, Al Basrah, Iraq's second largest city, was a ghost town. Crowds of people had been replaced by heaps of sandbags, and the normally crowded open marketplaces were hardly recognizable.

The Japanese representative came to the customhouse to prove that his company

*Attendance Fee: When the client wishes to appoint specific subcontractors for some parts of the construction work, such as electrical and mechanical installation and piling, the main contractor will become responsible for the work of the nominated subcontractors and thus receive an attendance fee for supervising these subcontractors.

had properly shipped scaffolding items to Kuwait. It had been charged that the items were different from those on the import list filed earlier for use on a construction project, and, as a result, a trailer truck was taken into custody by the Safwan customhouse at the border with Kuwait, with the truck driver also in custody. The Japanese representative submitted a petition that had been prepared in close consultation with the accompanying Iraqi lawyer. Thanks to this effort, the Iraqi customs manager in charge of the case levied only a small fine and allowed the trailer truck and the driver to cross the border. Upon the settlement of that case, the official in the customhouse pointed to three files of paper piled haphazardly on his desk saying these were for 150 court cases under contention, a reminder of how important it is to properly handle customs-related matters in Iraq.

Iraq enforces a unique customs system. Although a tariff law and tariff rate laws exist, the legal contents are freely interpreted by the officer in charge. This causes one problem after another, since most of the necessary construction machinery and materials must be brought into Iraq from Japan or a third country, where local supplies of these goods are short. In particular, those construction materials, machinery, and vehicles imported as temporary items must be reexported, donated to the customs office, or transferred to another project in Iraq after the completion of the project concerned. Without proper after-completion processing, the initial import guarantee pledged by the contractor cannot be satisfied, making the project incomplete in the minds of Iraqi officials.

Yet the after-completion processing of temporary items is a source of constant problems stemming from processing mistakes, documentation errors, and differences in the interpretations of the laws by customs officials and contractors. Almost all foreign contractors, including those from Japan, have been forced to pay large penalties, and some contractors have paid fines amounting to millions of dollars or even tens of millions of dollars. In addition to customs processing, Iraq provides an abundance of war-related risks, and contractors are compelled to upgrade their project management capability to cope with the wide variety of risks present in that part of the world.

Risk Management

All people and organizations are traveling through a time tunnel exiting into the twenty-first century. In all likelihood, the supplies of natural resources, energy, construction space, and funds will dwindle. Because of this, conflicts among corporations or governments will increase and magnify risks associated with specific projects and specific countries. Although some of these risks can be effectively avoided by improving the contractor's capability as a system organizer and upgrading the quality of each employee assigned to overseas projects, many large risks remain as a great obstacle to the expansion of international operations. The methods of risk management adopted by other outbound industries are not necessarily effective for construction companies, because construction is essentially a made-to-order trade with a relatively long period required to complete a product, from two to three years to as long as 10 years. Contractors will have to develop their own risk management methods that will suit the characteristics of the construction trade.

Nevertheless, Japanese contractors are relatively inexperienced in risk management, lack reliable widely applicable theories in risk management, and are attempting to minimize risks by trial and error at present. Even if risks are apparent, contractors tend to overlook the seriousness of the surrounding risks if they do not suffer actual losses or damage. To reach a high level of risk management, a plenitude of practical experience and an in-depth analysis of each type of risk will be necessary, and case studies of medium- and long-range risk management activities of both Japanese and foreign contractors must be collected for close examination.

Only a few Japanese contractors have started to make a full analysis and assessment of local construction environments in the country concerned in order to draw up contingency plans, but Japanese contractors as a whole are beginning to realize the importance of accurately evaluating contingency factors when making a cost estimate for each project. In the past, the general rule was "the greater the risk, the bigger the profit," but present high-risk projects may show only a mediocre profit. Risk management, including contingency assessment, has become much more important.

Risks surrounding a project may be divided into three groups according to cause—emergency, external, and internal risks (Table 3.3). Emergency risks are so-called *country risks*, and involve dramatic changes in the economic, financial, or social conditions of the host country due to the eruption of grave emergencies—such as the Iranian Revolution, the Iran-Iraq War, and the Cambodian civil war—which can cause the contractor to sustain a loss or gain only an unexpectedly small profit from the project. External risks are brought about by less dramatic changes in policies, economic conditions, natural environments, and other factors that cannot be controlled by the contractor. Internal risks, on the other hand, are caused by mistakes and failures made by the contractor, including delays in work schedules caused by careless planning, incomplete preparation of contract and specification documents, misjudgment of various data, and inability to deal with a client's claims.

Although risk management is a difficult task, one systematic approach to the problem of risks is to convert risk experiences into easily retrievable and analyzable data. A computer processing of this data will provide useful information for

Table 3.3 Risk Factors Involved in Overseas Projects

1. *Emergency Risks*
   - War (wars, hostilities, revolutions, civil wars, revolts)
   - Confiscation (nationalization, creeping confiscation)
   - Restricted remittance (incapability, restriction, prohibition, and cessation concerning remittance)
2. *External Risks*
   - Frequent changes in laws and regulations by local government
   - Chronic deficits in balance of international payments and chronic shortage of foreign currencies suffered by the host country
   - Natural environment and geological conditions
   - Local partners and agents (gradual increase in equity share and management decision of local partners)
   - Clients (failure to pay for liabilities)
   - Changes in foreign exchange rates and estimated costs
   - Inflation
   - Illegal confiscation of bonds
3. *Internal Risks*
   - Work Processes
   - Control of labor and subcontractors
   - Procurement of machinery and materials
   - Contract documentation and specifications
   - Organization of joint ventures, consortiums, etc.
   - Defective planning (inaccurate feasibility study)
   - Inappropriate treatment of client's complaints

determining measures to minimize risks or to deal effectively with damage incurred. Further, to realize an efficient cost management capability, Japanese contractors are expected to develop computer programs for drawing up contingency plans.

Protection from Risks

Both private and government insurance is available against risks concerning overseas construction projects. Private insurance covers most of the risks, and government insurance provides supplementary coverage on the exceptional types of risks not protected by private insurance companies. The main Japanese government insurance is *technology provision insurance* which protects the companies against uncollectible credits on exports of technologies due to occurrences that are the responsibility of the client, such as business failures, inability to pay, and breach of contract stipulations on the part of the client. Technology provision insurance also gives protection against losses and damage caused by unforeseeable developments, and contractors are insured against noncompletion of projects due to the eruption of a war, for example. But there are a number of risks still not covered by insurance, and contractors must prompt the government to initiate tax measures to minimize overseas risks, such as the creation of a tax-exempt reserve for overseas construction operations.

INTERNATIONAL EXPANSION

As the amount of construction orders from abroad has flattened out over the past three or four years, after a decade of brisk growth, it is time for Japanese contractors to review their market approaches. The construction orders received to date fall into the following categories:

- Projects funded by yen loans or grants from the Japanese government
- Projects funded by loans from the World Bank or the Asian Development Bank
- Projects obtained through international tenders, on the basis of information provided by Japanese trading houses and industrial plant exporters
- Projects obtained directly from local clients by the contractor's overseas branch offices
- Projects obtained from Japanese clients planning to build business facilities in foreign countries

To expand overseas work, Japanese contractors must improve these approaches to the international market and upgrade their information-gathering and marketing capabilities to become competent international system organizers. To this end, five strategies may be applied:

1. Demand creation
2. Operation expansion
3. Multinationalization
4. Technological progress
5. International financing

Creating Demand

In early January 1985, a long line of London policemen fenced off citizens from the Piccadilly Street and Regent Street quarter, and the subway line running through the area was temporarily stopped. All this was in the aftermath of a building explosion that claimed 10 casualties and was triggered by a city gas leakage. Similar accidents are reportedly frequent in London because of the city's aged buildings and gas piping, and a number of Japanese construction companies are now taking orders for renewal work in Britain. In the United States, Australia, and elsewhere, many large and medium contractors from Japan have invested in real estate development projects, hotels, office buildings, and shopping and leisure facilities. Thanks to the growing trade surplus in Japan and the lower costs of foreign loans, investments by Japanese contractors are particularly noticeable on the Australian east coast, Southeast Asia, the U. S. west coast, and China. It has been predicted that the economic center of the world will shift from the Atlantic to the Pacific Basin by the twenty-first Century, and Japanese contractors foresee a growth and increase in diverse demand around major Pacific rim cities, such as Sydney, Melbourne, Auckland, Los Angeles, San Francisco, Bangkok, Singapore, Hong Kong, Peking, Shanghai, and Dairen.

BUILDING THE THAI RIVIERA. Now enjoying worldwide fame as a tropical resort, Pattaya—about 93 miles south of Bangkok—was a nameless fishing hamlet until a little over 10 years ago. Since the discovery of this secluded community by an entrepreneurial Frenchman, hotels and resort houses have mushroomed along the shores of the Pattaya Riviera on the Gulf of Thailand, attracting hordes of vacationers from Europe, Japan, Hong Kong, Taiwan, and Australia. Pattaya, now a major dollar collector for Thailand, would have continued its hamlet existence of the past hundreds of years if it were not for that Frenchman, who tried to create new demand among holiday-loving people all over the world.

The success of Japanese contractors in future overseas operations will hinge, to a large extent, on how effectively they can stir up dormant demand. This will require an ability to sense potential demand from marketing activities and an ability to perform detailed market research and planning, as well as an ability to procure the necessary funds. The market scale and growth of host countries must be analyzed; the local political and economic policies must be taken into account; and market needs and clients' benefits should be anticipated and catered to from an early stage. The best after-care services must be offered, to help clients realize the most efficient use of their buildings. Further, contractors should increase their ability to assist clients in selling their products. For example, if the client is a hospital owner, the contractor should be able not only to procure, install, and operate the necessary medical equipment but also to recruit the necessary doctors. Or if the client is a physical fitness center operator, the contractor must be able to train or dispatch expert instructors. Tenants may be found for clients for commercial facilities, while tourist promotions may be pushed in the interest of clients in the leisure business.

KUMAGAI GUMI. A leading Zenecon, Kumagai Gumi is noted for creating demand through the establishment of joint ventures and participation in other corporations through shareholding. Kumagai Gumi took the order for a redevelopment of the old Madison Square Garden in New York in November 1985, for which the Japanese general contractor formed a redevelopment joint venture firm including its U.S. subsidiary and a local U.S. contractor. In Hawaii, Kumagai Gumi has announced that it will invest 600 billion yen ($4 billion) in a Ko Olina resort project as the developer, as well as constructor, of a major resort spot as glittering as Waikiki. By becoming a project entrepreneur, Kumagai Gumi apparently intends to gain a wide project experience, from development to after-care. Although Kumagai Gumi is very much an exception in its advanced international operations, the company's integrated activities—from development of potential demand to proposal, planning, design, funding, building, maintenance, and after-care—are regarded as the right formula for all internationally minded Japanese contractors. Today, contractors are becoming aware that they will not be able just to take orders, but must create a demand.

Improving Project Management

How do Japanese contractors see themselves in comparison with their counterparts in the West and newly industrialized countries? A survey on this question was carried out by the Nomura Research Institute in 1984 (Table 3.4). The results indicate that Japanese contractors considered themselves as being more advanced than their foreign counterparts, primarily in "hardware" capabilities: for example, punctuality in meeting completion schedules, quality of construction work, and flexible compliance of technologies to clients' needs—a natural result of serving finicky Japanese clients in the domestic market. On the other hand, the Japanese contractors felt that they lagged behind overseas companies primarily in "software" capabilities: for example, negotiation, dealing with complaints, foreign exchange operation, information collection, market research and planning, and project management. The results of the survey suggested that improvements by Japanese contractors in project management and overseas employee training have not kept pace with the rapid growth of overseas orders.

A method of enhancing project management would be to procure as much machinery, vehicles, and materials as possible from the host countries or a third

Table 3.4 Strong and Weak Points of Japanese Contractors—Self-Evaluation

	Comparison with					
	Industrial Countries[a]			Newly Industrializing Countries[a]		
	Strong	Equal	Weak	Strong	Equal	Weak
1. Information gathering, research and planning	2.3	31.8	65.9	28.6	54.8	16.7
2. Engineering construction ability	7.1	38.1	54.8	57.5	37.5	5.0
3. Project management	4.7	39.5	55.8	42.9	40.5	16.7
4. Abilities of dispatched employees	7.0	55.8	37.2	59.5	33.3	7.1
5. Procurement of machinery and materials	4.7	65.1	30.2	42.9	42.9	14.3
6. Collaboration with other industries	11.6	51.2	37.2	28.6	47.6	23.8
7. Construction and civil engineering technologies	38.1	52.4	9.5	82.9	9.8	7.3
8. Technological adaptability to host countries	20.9	69.8	9.3	57.1	38.1	4.8
9. Quality of construction work	72.7	27.3	0.0	93.0	7.0	0.0
10. Punctuality in completion of schedules	79.5	20.5	0.0	90.7	7.0	2.3
11. Negotiation and handling of complaints	0.0	6.8	93.2	4.8	50.0	45.2
12. Contribution to local community[b]	7.0	48.3	44.2	24.4	41.5	34.1
13. Local labor management and education	4.7	58.1	37.2	28.6	64.3	7.1
14. Reduction of labor costs	7.0	60.5	32.6	7.3	26.8	65.9
15. Reduction of costs of dispatched employees	16.3	53.5	30.2	9.8	19.5	70.7
16. Adaptation to local society and culture	2.3	37.2	60.5	12.2	58.5	29.3
17. Building of friendly human relationships	6.8	45.5	47.7	20.9	51.2	27.9
18. Procurement of necessary funds	18.6	44.2	37.2	61.9	38.1	0.0

Table 3.4 *(continued)*

	Comparison with					
	Industrial Countries[a]			Newly Industrializing Countries[a]		
	Strong	Equal	Weak	Strong	Equal	Weak
19. Determination of payment conditions	4.8	57.1	38.1	24.4	63.4	12.2
20. Accuracy in cost estimation	13.6	65.9	20.5	40.5	52.4	7.1
21. Hedge against foreign exchange loss	0.0	30.2	69.8	11.9	71.4	16.7
22. Protection against country risks	0.0	30.2	69.8	14.6	61.0	24.4
23. Support by home government	2.3	13.6	84.1	9.8	14.6	75.6
24. Overall evaluation	0.0	60.5	39.5	38.1	52.4	9.5

[a] Multiple choice; figures in percentage

[b] *Contribution to local community* includes transfer of technologies, use of local workers, and procurement of local machinery and materials.

Source: October 1984 survey by Nomura Research Institute at request by OCAJI. Respondents are Japanese construction companies with membership in the Overseas Construction Association of Japan, Inc. (OCAJI).

country to reduce both the procurement costs and the transport period. To this end, the establishment of regional centers across the world is promising. Some European contractors have deployed a representative office as a regional center in Istanbul, Turkey—a strategic location overlooking Libya, Egypt, Syria, Persian Gulf countries, and Iran—primarily to increase procurement efficiency.

Another method of improving project management is to hire an increased number of foreign people as managers, specialists, clerical staff, and labor. In this regard, the nationalities of those already recruited indicate a regional expansion from India, Pakistan, Bangladesh, China, the Philippines, Thailand, and Sri Lanka to the United States, Europe, Turkey, and Eastern Europe. The need for training facilities for these people is growing. The regional centers would be responsible for securing the supply of both goods and labor, making full use of computer data on procurement prices, suppliers, subcontractors, and job seekers in each region of the world.

Along with an increase in number, overseas orders will also show a change in content. Although construction work will continue to be the business mainstay in overseas orders, the proportion of nonconstruction work will significantly expand due to the growing importance of proposing and developing projects. Accordingly, contractors will be required to improve their research, planning, consulting, R&D, design, legal, financing, and after-care service capabilities in addition to the conventional function of administering construction work. The proportion of construction management contracts will increase, whereby the contractor will participate in projects as an adviser on the selection of a building contractor, the determination of contract terms, the management of construction work, and the procurement of the necessary funds. Simultaneously, the construction company may separate some of its departments into discrete corporations, to better implement the necessary diversification of its overseas operations. Forming transnational consortia with contractors, architectural designers, consultants, developers, banks, medical institutes, realtors, hotels, and many other types of businesses in the United States, Europe, and South Korea will become an everyday affair.

Becoming Multinational

An important way to create new demand and expand business operations is to go multinational. Today, however, many countries enforce tighter restrictions on the activities of foreign contractors. They have strengthened their requirements for foreign contractors to form joint ventures with local companies, have reduced international tenders due to a widespread economic slowdown, and harbor greater hopes of receiving construction technology transfers and various other contributions from foreign contractors to the local economy. To meet these demands, a Japanese contractor must be considered an insider by the host country, not as a contractor from the "outside." Specifically, contractors should buttress their locally incorporated subsidiaries and affiliates on a long-term basis, while simultaneously preserving the characteristics unique to each Japanese contractor.

According to the corporate multinationalization theory of Professor Noritake Kobayashi at Keio University in Tokyo, Japanese contractors are now in the second stage of a five-stage development of international corporate management (see Figure 3.3). The first stage involves an unsophisticated overseas management in which overseas operations are managed by the main office as if they were domestic operations. Presently in the middle of the second stage, Japanese contractors are beginning to shift some of the management responsibilities of the main office to their overseas subsidiaries and affiliates in response to the expectations of host countries, such as requests to form joint ventures with local firms.

The contractors must now move toward the third stage, during which they will open regional centers in the major cities of investment host countries to achieve an efficient, systematic utilization of labor, office staff, machinery, materials, funds,

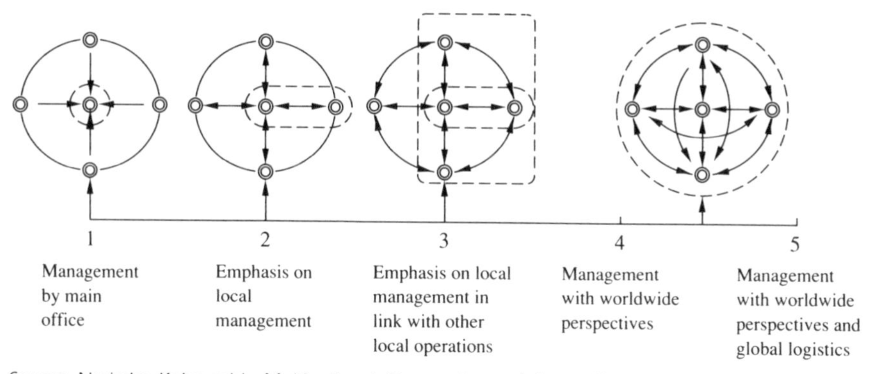

Source: Noritake Kobayashi, *Multinational Corporations of Japan* (*Nihon-no Takokuseki Kigyo'o*) (Tokyo: Chuo Keizaisha, 1980).

Figure 3.3 Transnationalization stages of construction operations

and management resources in different regions of the world. These key cities will include New York, Los Angeles, Sydney, Peking, Shanghai, Hong Kong, Istanbul, Cairo, Kuwait, Abu Dhabi, New Delhi, Singapore, and Rio de Janeiro.

Once their regional centers are deployed, perhaps from the 1990s into the next century, Japanese contractors will develop their markets around these regional centers to encompass surrounding areas in the fourth stage of international operations. When these market areas become large enough to overlap with each other, it will become possible, as the final stage, to implement global management strategies and to evolve from a regional into a mature international system organizer with an excellent logistics ability.

The international operations of a Japanese general contractor in the twenty-first century may have an organization as shown conceptually in Figure 3.4. With the operational headquarters located in Japan, the world market will be divided into eight major blocs of more or less equal importance: Japan, North America, Latin America, the Middle East and Africa, Southeast Asia, Oceania, and Europe. Within each bloc will function a number of regional centers, and there will be close, instant communication between the Japanese headquarters and the regional centers and between one regional center and another. This type of organization must be built on the basis of effective international strategies, enabling Japanese contractors to grow into efficient multinational corporations, such as IBM, ITT, Exxon, Philips, and Siemens.

Japanese contractors already harbor a strong desire to set up regional centers in China, due to the growing Chinese demand for modern hotels, office buildings,

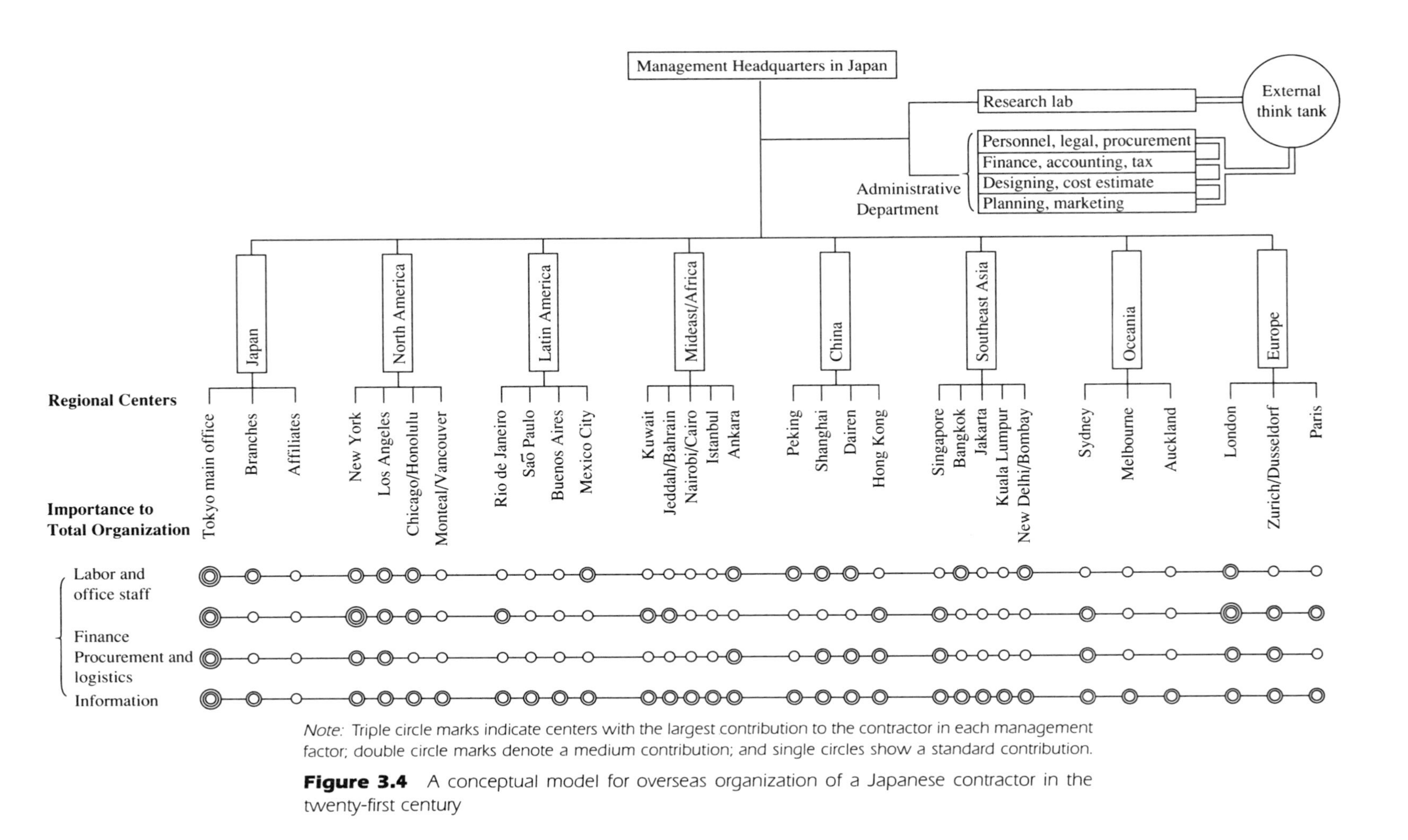

Note: Triple circle marks indicate centers with the largest contribution to the contractor in each management factor; double circle marks denote a medium contribution; and single circles show a standard contribution.

Figure 3.4 A conceptual model for overseas organization of a Japanese contractor in the twenty-first century

golf courses, and super–high-rise buildings. As a result, Shimizu Corporation and Takenaka Corporation have already formed joint ventures with Chinese partners, taking advantage of the 1979 China–foreign joint capital corporation law applicable to Chinese corporations with a foreign equity share of 25% or more. This law stipulates requirements for the transfer of advanced technologies and permits the remittance of income from China.

China is Japan's largest Official Development Assistance (ODA) beneficiary, accounting for a sizable 15% of Japanese monetary assistance to developing countries—a good sign in that China has the money to fund construction projects. Further, the country is expected to receive increasing assistance from international organizations in the future. As a result, Japanese contractors have high hopes for the Chinese market, although some Japanese investors from different industries have suffered financial losses from unpredictable changes in Chinese political and economic policies and from delays in project progress due to a lack of legal backup in some investment activities.

China poses other problems to foreign contractors, such as a requirement to make payments in the local yuan currency, a similar requirement to pay labor expenses to a Chinese administrative corporation in a foreign currency, and strong competition from hard-working local contractors. But Japanese contractors are hopeful that China will establish additional Special Economic Districts (e.g., Shenzhen and Zhuhai at present) and Economic Development Districts (e.g., Shanghai, Tienchin) to permit freer economic activities and thus stimulate China's demand for energy-, traffic-, and communications-related structures in the medium and long term.

Technology Transfer Strategy

History has proven that cultural assets move from haves to have-nots. Nevertheless, this is not the case today; there is a widening North–South gap. This disparity blocks stable international economic relations, a balanced distribution of wealth, and a smooth economic growth of the world as a whole. The transfer of technologies from North to South has preceded slowly, and the voices calling for technology recipient countries to upgrade their personnel and organizations to accelerate technological transfers are becoming louder. Consequently, many developing countries have drawn up medium- and long-range plans to build academic cities furnished with high-level educational institutes, research laboratories, and training schools and to establish information- and technology-intensive cities accommodating high-tech companies and facilities. Developing countries will, therefore, increase their investments for these construction projects, and industrialized countries will support their efforts by boosting monetary assistance.

In the Japanese market also, plans are being made to construct intelligent buildings and "smart" commercial zones furnished with INS (information network systems), cable television, office automation equipment, and other sophisticated information systems. In addition, Japanese contractors are optimistic about the plans mapped out by local governments in Japan to build high-tech cities that will house research and development facilities and educational institutes in each prefecture. In connection with these plans, contractors are eager to promote their technologies to build not only "intelligent" buildings but also air-supported domes, high-tech clean rooms, rock oil tanks (unlined rock caverns for oil storage), and solar energy utilization facilities, among other structures. Japanese contractors are also likely to apply their high-tech approach to foreign markets in the near future.

Mosque-shaped, air-supported domes such as soccer stadiums may stir up strong demand in the Middle East and Latin America, where soccer is very popular. Clean rooms are good prospects in developing countries, where the use of computer systems and the number of medical facilities will gradually increase. This is particularly true for Mideastern and African countries whose manufacturing and medical productivity can be raised only by filtering out the dust from surrounding deserts. Although oil-producing countries have curbed their construction activities, they continue to show a strong interest in energy-related technologies and are willing to set aside special budgets for attractive energy projects. Anticipating that its inland oil wells will be exhausted within a few decades, Kuwait is building offshore oil rig platforms in the Persian Gulf, while some countries are trying to acquire oil-substitute energy technologies from industrialized countries. In Baghdad, Iraq, the solar-powered Solar Energy Research Institute has been running smoothly since its completion in 1984 by Shimizu Corporation, and along the banks of the Tigris in the Iraqi capital, a European contractor has built rows of flat apartment houses equipped with solar panels.

The desires of developing countries to acquire advanced technologies are far stronger than people in industrialized countries imagine. Contractors can capitalize on these desires by analyzing the need for high technologies in each developing country and collaborating with high-tech manufacturers to combine advanced technologies with construction. In the industrialized part of the world, new markets will also likely emerge. Financial centers such as New York, London, Hong Kong, Singapore, Sydney, Zurich, and Bahrain will require new buildings with greater information-processing and communications capabilities. Similar needs will arise in Pacific rim cities as the Pan-Pacific communications link grows stronger, giving Japanese contractors an advantage because of the availability of Japanese high technologies in communications. Advanced technologies, together with the excellent financial status enjoyed by Japan, will be a powerful tool for Japanese contractors in their bid to create new demand in the world market to expand their international business, and to evolve into multinational corporations.

INTERNATIONALIZING FROM INSIDE

When the need of Japanese contractors to become international corporations is discussed, questions usually arise over ways to expand operations abroad. But an important factor is that Japanese society is rapidly internationalizing itself. The number of offices of foreign companies is increasing rapidly in Tokyo business quarters, and it is now easy to find people from the United States, Europe, other Asian countries, and Africa among Tokyo pedestrians. As Japan is going international from the inside, the Japanese construction industry cannot stay unaffected by this change. It must respond to the needs of incoming foreign companies and foreign representatives for specific types of offices, factories, and housing. These prospective clients from abroad have different construction needs because of their different life-styles and business practices. Japanese contractors can best serve foreign residents in Japan by applying their knowledge of world life-styles derived from overseas construction projects. It will also be necessary to supply foreign companies planning to enter Japan with information on land prices, environmental conditions, and other practical information on Japan.

As another aspect of the internationalizing trend inside Japan, an increasing number of foreign contractors and engineering firms have opened offices in Japan, and with the New Kansai International Airport, Trans-Tokyo Bay Highway, and other big projects on the drawing board, the Japanese construction market is looking attractive to contractors from the United States, Europe, and South Korea. In the not-too-distant future, Japanese contractors will be competing with foreign members of the domestic market, which will in turn prompt the Japanese to become more international, while at the same time many of the incoming foreign contractors will become familiar with the Japanese ways of doing business.

The third internationalizing factor is the increasing influx of foreign construction workers into Japan. Supplies of both skilled and unskilled workers, engineers, and legal experts from other countries will increase. Especially construction people in neighboring Asian countries will regard Japan as an alluring, high-wage employment market. Although most countries regulate labor movements across national borders, the ongoing migration of work force into richer countries cannot be totally stopped. Since Japanese contractors may experience an acute labor shortage in the future due to the aging of the Japanese population, they may require foreign workers in remarkably larger numbers than at present.

These three types of internationalizing forces in the Japanese market will affect not only leading general contractors but also small regional contractors and subcontractors. The Japanese construction industry requires such grass-roots internationalization to achieve truly global operations.

AIMING AT THE WORLD'S SUPER PROJECTS

The growing interdependence of national economies has caused widespread trade protectionist moves and worldwide recessions, but on the brighter side it has also brought the possibility of organizing large international construction projects. In 1977, Mitsubishi Research Institute, led by then-president Masaki Nakajima, proposed the Global Infrastructure Fund (GIF) program, intended to overcome world recessions and the North–South disparities by implementing large-scale international construction projects using mainly the funds and technologies of industrialized countries. GIF advocates hope to collect a $500 billion fund from industrialized, oil-producing, and socialist countries to plan and run super projects that will

1. Help the world to save natural resources and energy and expand food output
2. Smooth the healthy flow of international finances
3. Promote technical innovations
4. Enhance peaceful relationships among countries

The Japanese think tank has proposed dozens of construction projects to be undertaken throughout the world, including an undersea railway and highway tunnel across the Strait of Gibraltar, a water-power generation system in the Himalayas, a sea-to-sea canal across the Isthmus of Kra in the northern Malay Peninsula, and a current control dam in the Bering Strait.

The plan to build a second canal in Panama looms as the most realizable GIF project as of today. This plan calls for the construction of a Pacific–Atlantic canal 61 miles long (36 miles inland and 25 miles offshore), 108 feet deep, 656 feet wide inland, and 1,312 feet wide offshore. This massive canal, when completed after a 10-year project period and a total spending of $20 billion, will enable the passage of 500,000 DWT* vessels at high tide and 300,000 DWT at low tide, producing stimulant ripple effects on the world economy. The plan was initially conceived by the United States and Panama but later a Japanese industrial group joined, led

* Dead Weight Tonnage (DWT) is the total weight in British tons of all possible cargoes, passengers, fuels, foodstuffs that can be loaded in a ship.

by Mitsubishi Corporation and with a membership comprising Penta-Ocean Construction, Shimizu Corporation, Kajima, Taisei, Mitsubishi Heavy Industries, the Bank of Tokyo, and the Industrial Bank of Japan. Japanese contractors are eager to take part in similar projects of worldwide significance, in order to take advantage of their advanced construction technologies and expanding system-organizing capabilities.

Hamamatsucho Toshiba Building and Tokyo Gas Building (Japan) Shimizu Corporation

CHAPTER 4

NEW BUSINESS DEVELOPMENT STRATEGY

LAUNCHING NEW BUSINESS

A Time of Change

The life cycle of a corporation, from growth to maturity to aging, is said to last 30 to 50 years, with an accelerating tendency for this cycle to become shorter and shorter. Modern business history shows that only those companies capable of adapting to the changes caused by time have survived. Chairman Kiichi Noji of Shimizu Corporation has said, "'Maintenance of the status quo is a dead end.'" Yet the current condition of the construction market makes it hard to maintain even the status quo. Contractors today may be caricatured as children trying to climb the "down" escalator. If the children climb as fast as the steps descend, they will stay in the same position; if they want to go higher, then they must climb faster than the speed of descent. Every time they take a breath, they move down a few steps.

The present time, though basically hard for contractors, also offer great business opportunities because the market is rapidly changing. To cash in on market changes, it will be necessary for Japanese contractors to look back on their past achievements and map out future images of themselves, with emphasis on the comprehensive activity and frontier spirit ingrained in the construction business.

Concepts of New Business

The growth strategy for contractors may be divided into four approaches, as shown in Figure 4.1:

A. Penetration of the existing market
B. Development of new technologies
C. Development of new market segments
D. Diversification into new business lines

Approach A involves an effort to garner a greater number of orders by, for example, increasing sales staff and bolstering the price competitiveness through increased productivity. Approaches B and C involve the adding of new construction business lines such as engineering construction, overseas construction, developer business, and technological development—all of which involve areas associated with construction. This chapter focuses on Approach D, which deals with business diversification only to nonconstruction business. Note that since usually two or more approaches are taken simultaneously, their interactional effects should not be overlooked.

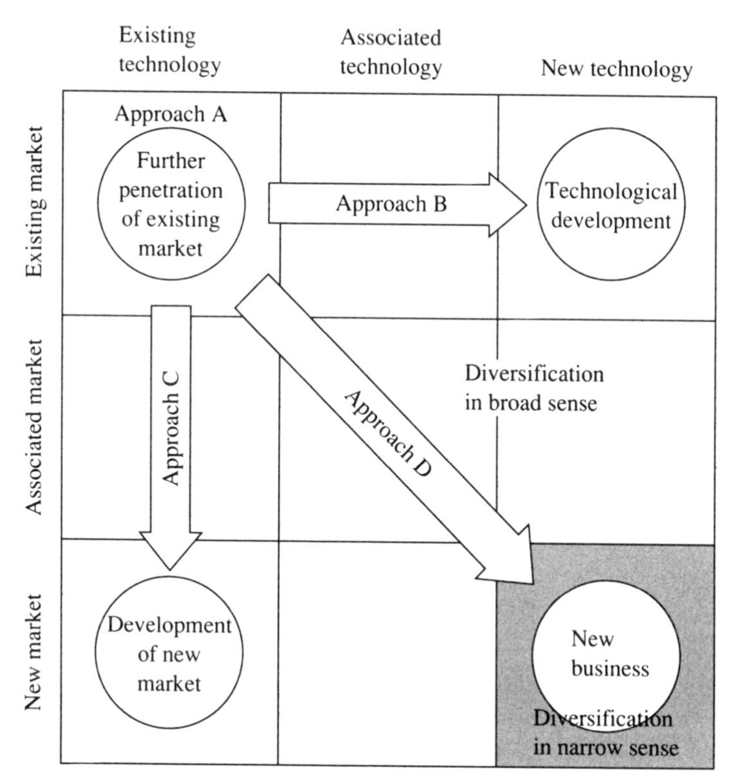

Figure 4.1 Concept of new business

TYPES OF NEW BUSINESS

Aims of New Business

Starting new nonconstruction business appears to be popular among Japanese contractors (see Table 4.1). Reflecting future growth of the information and other service industries, a large number of contractors have entered this field, especially the sales of construction database services and construction software for CAD and small business computer use. Another popular area is the sports and leisure business, including hotel and tennis club operations. Both the information and leisure ventures are aimed at an efficient utilization of existing management resources, such as unused real estate properties, personnel, and technologies. In addition, there are some instances where totally new business has been attempted, such as the operation of a restaurant in Guam and sales of educational video films for prep school students.

Table 4.1 New Business Recently Started by Leading Contractors

Area	Kajima	Taisei	Takenaka
Information and communication	• Sales of construction software	• Sales of construction software	• Sales of construction software • Communications service for tenants
Housing	• Interior business	• Sales of unitized furniture	• Establishment of reform centers
Sports	• Tennis club operation	• Sports club operation	• Golf course operation
Lesiure	• Hotel operation	• Leisure business • Hotel operation	• Hotel operation
Related to construction	• Development of carbon fiber reinforced concrete • Consultant business	• Development of carbon fiber reinforced concrete • Development of construction materials for hobby use • Consultant business	
Others	• Leasing • Publishing	• Leasing	• Leasing

A common aim of all business diversification efforts is to increase the number of income sources, at a time when the price competition is suppressing the profit margins of construction orders to a painful level. Contractors are hopeful of protecting their profit performances by securing additional income sources. Another aim is to secure business stability and growth. This involves diversification into peripheral construction businesses (i.e., engineering construction, international

Ohbayashi	Shimizu	Others
• Sales of construction software • Sales & leasing of computers	• Sales of construction software • Communications service for tenants	• Sales of restaurant POS[a] • Sales of construction CAD systems
• Reform business	• Reform business	• Reform business
• Golf course operation	• Tennis club operation	• Sports club operation
		• Resort development • Hotel operation
• Production of flame retardant materials • Consultant business	• Development of carbon fiber reinforced concrete • Consultant business	
• Restaurant operation	• Leasing • Health care for the elderly	• Restaurant operation • Sales of educational video films • Overseas gold mining • Information business school operation

[a]POS = point-of-sale computer systems.

construction projects, real estate development) or nonconstruction business, and both usually have a synergistic effect of stimulating orders for construction work. The third aim is to sell in-house technologies and know-how as a commercial product. By doing so, the contractor may recoup some research and development investments, implant a cost-and-profit concept among the research and development staff, gradually transform the technology department into a profit-making machine, or find out about the practical performance of the in-house technologies through feedback from the users.

Another aim of starting new businesses is to provide employees with practical opportunities to learn management and marketing techniques in different specialized lines of business. This objective is best achieved by starting partnerships with nonconstruction companies. Other objectives include full utilization of current personnel, financial, and real estate resources; saving of costs by replacing outside agents and subcontractors with spin-off groups; stimulation of employees and the company organization by introducing brand-new business; and utilization of new business ventures as an antenna to sense the growing needs of consumers and users. These aims are mutually supplementary and interactional.

Case History of Kajima

A series of articles in a leading Japanese economic newspaper in 1985 introduced Kajima's management strategies for the future,* and the report proved of great interest to the members of the construction industry. According to the articles, Kajima, a top general contractor, ambitiously projects an average growth of 7.5% per year in the company's total sales from 1984 to 2001, far surpassing the projected GNP growth. But since the predictable growth of public works is weak, Kajima mapped out three strategies to expand sales without depending heavily on public works.

The first strategy is to increase market share in the private construction market. This will be achieved by a bolstered engineering construction capability, which will enable Kajima to win private orders in full packages including planning, design, construction work, mechanical and electrical procurement and installation, and after-completion maintenance services. In addition, Kajima plans to expand the development business in order to create demand. It expects that construction orders will be increased also through the development and sales of real estate products. The second strategy is aimed at the expansion of overseas construction work to 20 to 30% of the company's total sales, a minimum requirement to evolve into a truly international corporation. Kajima computes that this will necessitate a growth of 16 to 19% per year in overseas sales, much higher than the past growth rate, for the remainder of this century. The third strategy focuses on the cultivation of new business lines, and Kajima hopes to earn eventually 20 to 30% of total sales from nonconstruction operations (see Figure 4.2).

Assuming a long-term growth of 7.5% per year, Kajima's total sales will reach 3.2 trillion yen ($21 billion) in the year 2001, including 960 billion yen ($6.4 billion) sales from new business or 30% of total sales (see Figure 4.2). In 1984

*Nikkei Sangyo Shinbun, *Case History of Kajima* (Tokyo: Nihon Keizai Shinbunsha, July 7-10, 1985).

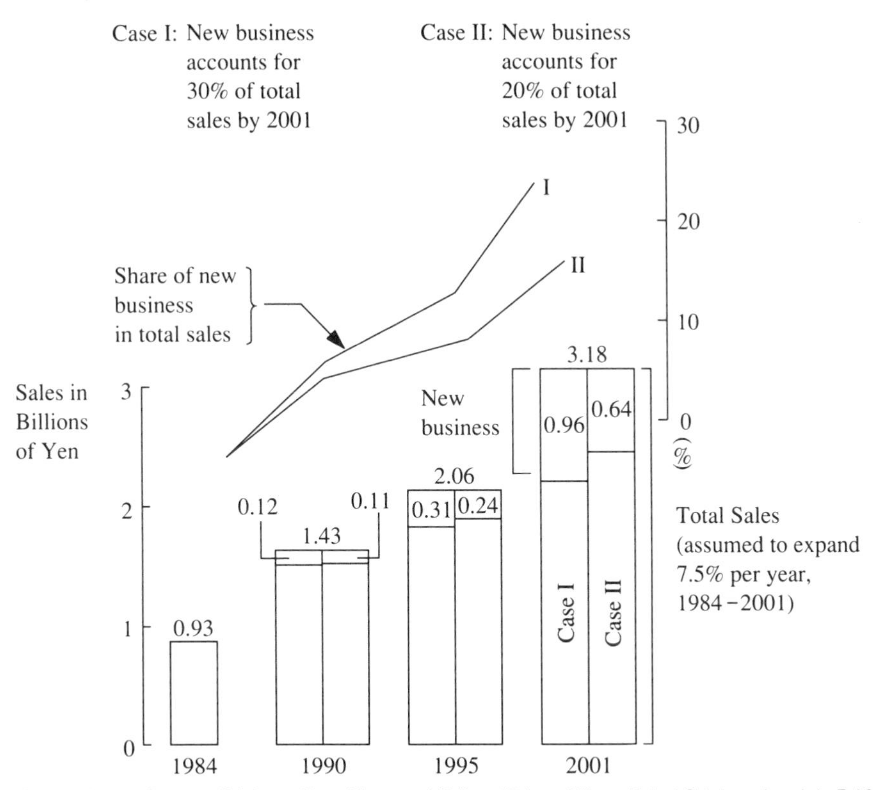

Source: Nikkei Sangyo Shinbun, *Case History of Kajima* (Tokyo: Nihon Keizai Shinbunsha, July 7-10, 1985).

Figure 4.2 Projected sales by Kajima Corporation

Kajima posted total annual sales of 930 billion yen ($6.2 billion), so that an amount of sales equal to the current total sales of the company will be necessary by 2001, only 16 years from 1985, just to achieve the projected nonconstruction sales—indeed an ambitious projection. Since 1984 Kajima has set up three subsidiaries specializing in new business, and their combined sales must be expanded by 20% every year from the current level of 50 billion yen ($0.3 billion) a year to ring up the targeted nonconstruction sales for the year 2001.

Kajima also provides for less ambitious earnings of 20%, instead of 30%, of total sales from new business by the year 2001. In this case, the targeted sales from new business will be 640 billion yen ($4.2 billion)—about the same turnover as Japan's largest textile producer, Toray Industries—requiring a 17% growth in new business every year until 2001. Kajima, therefore, will have to rapidly expand its new nonconstruction business, whether the target is 30% or 20% of total sales.

Kajima's new business strategy is notable for its ambitious targeting. Nevertheless, it is based on a cautious approach of developing new business closely related to the construction mainstay, businesses that will help stimulate construction orders; the company has abandoned the past tendency of Japanese general contractors to embark on any new business that can be easily started. Kajima makes it a rule to let the employee who has made a proposal for new business run the subsidiary initiated for this new business. Since the new leader is highly motivated, there is a greater chance of succeeding in the new venture. This personnel policy also motivates the employees in the main office to think constantly of new business that they may want to engage in. Kajima's experiments are being attempted by other large general contractors to various extents.

ELEMENTS OF SUCCESS

On the whole, new business development by contractors is still in the launching stage, and the importance of diversification efforts is increasing in the light of a grim outlook for the main construction business. Yet new business provides not only new opportunities but also new risks. For this reason, contractors are compelled to minimize risks by trying to build new business ventures at the outskirts of the construction area, where their experiences can be best put to use. Success hinges largely on which types of new business the contractor will choose and whether the company will be able to reshape its organization to best suit the new business.

Ideas about New Business

An ad hoc team for new business exploration, formed by Ohbayashi, a large construction company, recently conducted a survey to determine the types of new business the Ohbayashi employees wished their company to start (see Figures 4.3 and 4.4). A similar in-house poll was carried out among Shimizu Corporation employees, and the results were similar to those obtained by Ohbayashi. In sum, both surveys found that the largest number of employees considered that new business should be closely related to the construction trade; the popular business fields outside construction were sports and leisure, general services, and the real estate business, in that order.

By age, the older employees were more in favor of new ventures linked with the construction mainstay. But the reason for the older employees' strong attachment to construction is undeterminable: Is it because they are heavily committed to construction because they are overconservative, or because they know where the

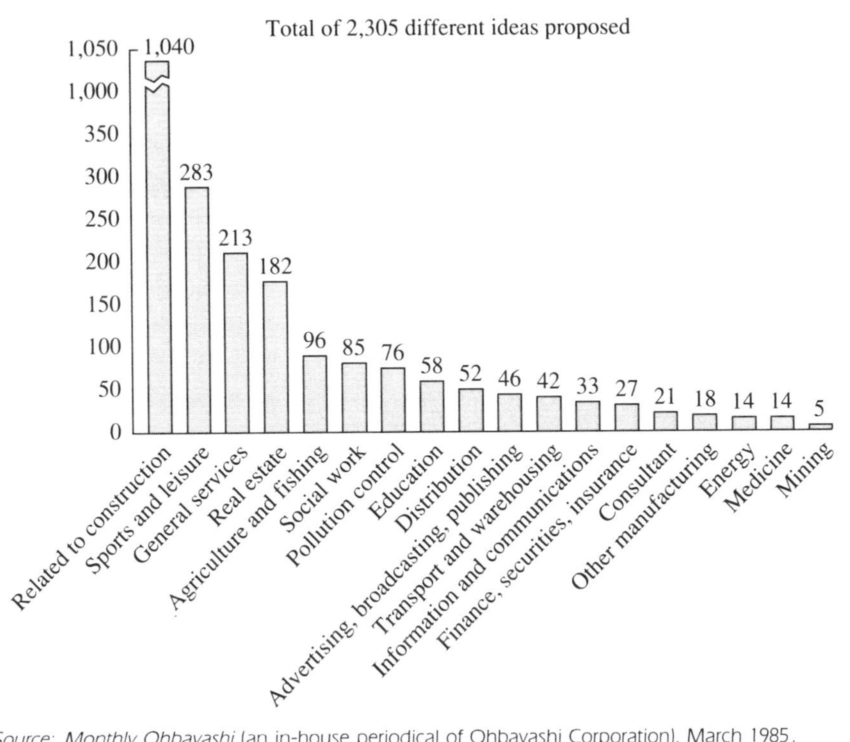

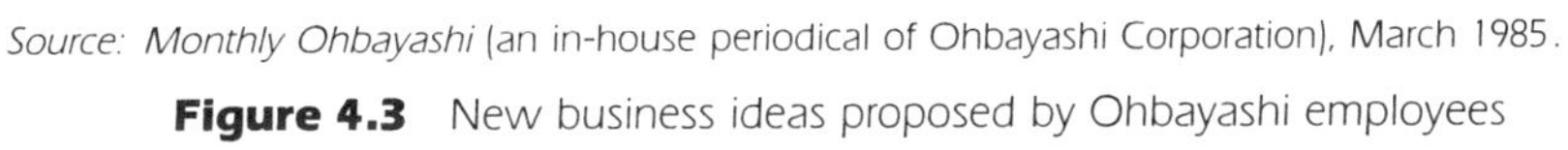
Source: Monthly Ohbayashi (an in-house periodical of Ohbayashi Corporation), March 1985.

Figure 4.3 New business ideas proposed by Ohbayashi employees

true strength of their company lies? Any new business that may flower in the next 10 years should already show some buds today, but it is important to perceive subtle but significant changes, to maintain enthusiasm for launching new business, and to teach the right ideas of combining proper available resources.

Keys to Success in New Business

Japanese contractors are only beginning to branch out into new business fields, and their know-how of how to succeed is still sadly lacking. Yet the experiences of companies in other industries suggest several key points for contractors wishing to succeed in new ventures.

ENTER NEW MARKETS WITH A GOOD GROWTH POTENTIAL. As a new economic power, Japan is experiencing a rapid industrial transformation. The change in the nation's industrial structure is determined by the dynamic interaction

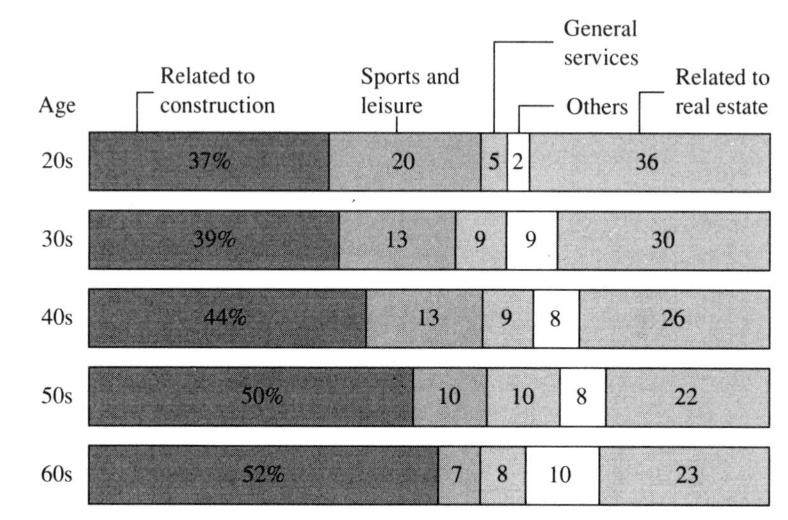

Source: Monthly Ohbayashi (an in-house periodical of Ohbayashi Corporation), March 1985.

Figure 4.4 New business ideas of Ohbayashi employees by age

of various growth-promoting factors and growth-inhibiting factors (see Figure 4.5), and the key to success is to find market areas with a maximum number of growth-promoting factors. Some of the key words for market areas considered to have a good growth potential are *culture*, *education*, *sports*, *leisure*, *urban redevelopment*, *medicine*, *health*, *information*, *communication*, *mechatronics*, *new materials*, and *biotechnology*. New business may be launched in these and other promising markets after an in-depth analysis of market growth potential, maximum market scales, management resources (e.g., available work force, talents, technologies, clientele, distribution channels), competitors, legal regulations, and the company's own strengths and weaknesses.

OBSERVE CHANGES AND BE FLEXIBLY ADAPTABLE. Human civilization may be described as the history of cross-encounters by different civilizations. People will gradually transform their own culture after coming in contact with a different culture. The experience of an alien culture has usually served as a prime mover for progress. Since prehistoric times, the Japanese have acquired vital cultural heritages from other cultures, such as bronze and iron ware, rice farming, Buddhism, Confucianism, firearms, and Christianity, mostly from the Asian continent. As did the earlier Japanese, the Japanese of today actively respond to the impinging influences of foreign cultures with a flexible mind. Corporations do not create new business or new ideas. New business and new ideas are generated when the political, economic, social, cultural, and technological environments change. This is a reason for the importance of mingling with companies from different industries when trying to start a new business.

CLARIFY BUSINESS GOALS. In attempting to start a new business, contractors should determine why they want to start the new business. If the answer is "because other companies have started the business" or "because we badly need a quick profit," then the contractors lack a needed viable goal.

CAPITALIZE ON YOUR STRENGTH. Any new business should be differentiated from that of competitors by capitalizing on the strong features of the contractor

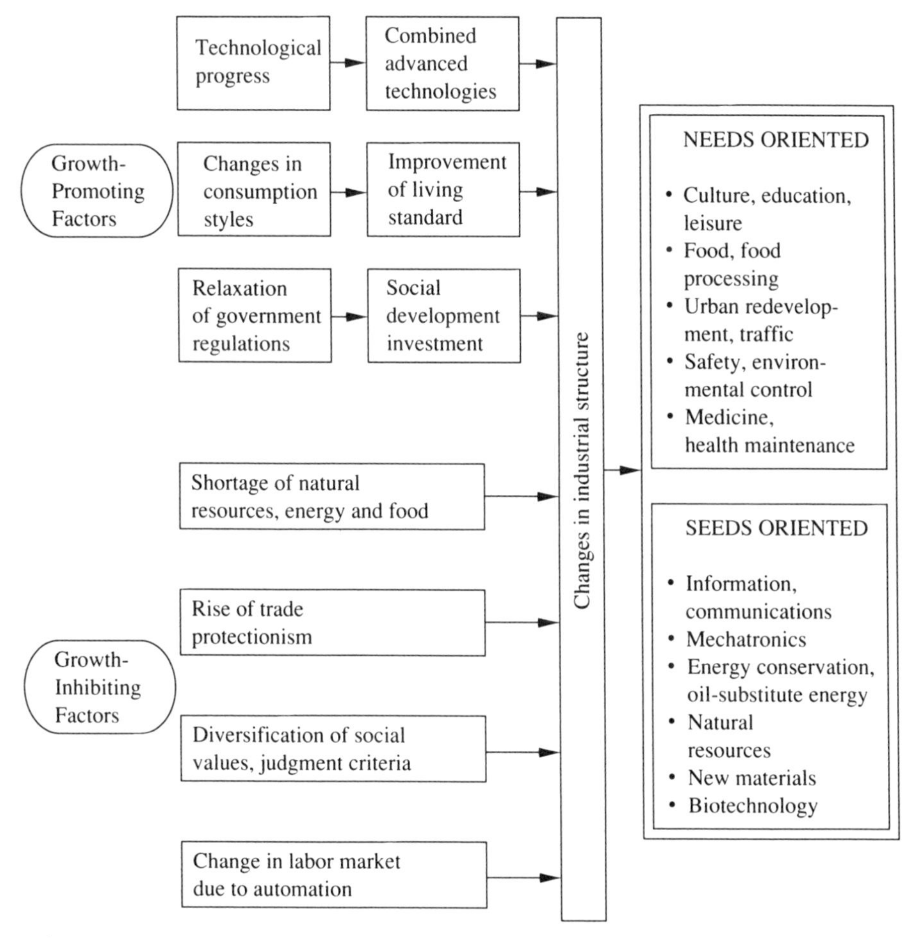

Figure 4.5 Factors promoting/inhibiting growth and market areas with a good growth potential

so that cutthroat competition can be avoided. Strength may be found in planning abilities, designing techniques, manufacturing technologies, material development, relationships with clients, distribution channels, personnel quality, organizational efficiency, administrative know-how, construction expertise, procurement capabilities, corporate images, and low prices, among other factors. General contractors have long been engaged in multifaceted operations in the construction business with little experience in specialized activities. A future task of contractors would be to make a specialized approach to new business, like a department store planning to open specialty stores.

AIM AT HIGHER VALUE-ADDED BUSINESS. One way to derive higher value-added business is to move *downstream* or to go *smart* in the river of merchandise. As a product moves from upstream to downstream, or from raw materials to semi-processed materials to parts to assembled product to systems to services, value is added at each stage. Going smart is raising the level of technological intensiveness; for example, improving quality, lowering prices, reducing the product size, and developing specialized and extreme technologies. A good example of a downstream success by an upstream company is the sale of popular ice cream making machines to consumers by Nippon Light Metal, Japan's largest aluminum smelter. A good example of going smart is the price reduction and capacity increase of electronic chips.

BALANCE THE EXISTING, PERIPHERAL, AND NEW BUSINESS. In the past, when the main construction business was expanding rapidly, contractors tried to increase their market share by increasing marketing staff and channels and by bolstering price competitiveness through improved production efficiency. But today, the main construction business has reached maturity, and the contractors are forced to look for other growth-oriented markets to achieve corporate growth. A market will grow because somebody has sown the seeds, and a share in such a growing market is determined by the size of research and development investments made by companies, in terms of the number of researchers and the amount of equipment allocated to foster the new business. Given the relatively large risks ingrained in new business ventures, the contractor must correctly relocate available business resources, so that a healthy balance between the main construction, peripheral, and new business lines can be achieved.

Japanese contractors have sometimes launched new business ventures on impulse, when construction orders dropped, but when the business entered a high phase, they tended to become impatient with the snail's pace of progress of their new business, cutting the amount of investment for the new venture. Continuity and consistency are vital ingredients in the new business recipe. The best approach is to aim at the construction peripheral field, where the main construction business and the burgeoning new business are expected to have a mutually beneficial relationship.

EMPHASIZE THE TIMING AND SPEED IN LAUNCHING NEW BUSINESS. Many new business ventures with large growth potential are considered to match the key words *timing*, *speed*, and *versatility*. A good example is the rise in the past several years of door-to-door package transport business by private trucking companies, which have taken a large share from the previously dominant Japanese National Railways,* a government railway company. The private package transport service is flourishing because if offers a faster delivery schedule and more convenient dispatch and receipt arrangements. As another example, McDonald's (Japan), a 50-50 venture between the United States and Japan, has grown into the largest catering business chain in Japan with annual sales of 100 billion yen ($710 million), thanks to its fast, convenient services. Similarly the Seven-Eleven Japan convenience store chain, a licensee of the U.S. corporation Southland, is rising as a major retail force in Japan, winning strong support from students and young working couples.

Industries, corporations, undertakings, and products all follow their own life cycles consisting of the budding, growth, maturity, and aging stages. They can also pass geographical stages starting from the industrialized country market to the industrializing country market to the developing country market. This geographical cycle may be followed smoothly by multinationalizing the company, by promoting a division of labor with foreign countries, and by making the best use of available overseas resources and tax regulations.

AVOID COMPETITION IN THE FULLEST SENSE. In the business community, a general rule is that market share and profit are proportionally related. Thus, a natural step to take in starting a new business venture is to aim at obtaining the largest share in a particular market, and there are only two ways of earning the leading share: (1) to enter the market before anybody else, or (2) to overtake the competitors who have entered the market before you. If equipped with the powerful distribution channels, production technologies, and funding capabilities of Toyota Motor (Japan's largest automaker) and Matsushita Electric (Japan's largest consumer electronics conglomerate), it would be relatively easy to be a latecomer and still catch up with the front runners. Other less gifted companies, however, must also try to move into the market before other companies; this is the best way to avoid competition, as pointed out many centuries ago by a Chinese strategist, Sun-tzu. This early entry will provide the company with the maximum chance of

*Because of the vast amount of deficit accumulated, the Japanese National Railways was split into six regional private railway companies in April 1987.

improving product quality and services and lowering prices beyond the practical reach of competitors.

Overcoming New Business Risks

New business ventures pose two types of risks to be reckoned with. The first type comprises risks arising from the nature of the new business itself, and the second type includes those stemming from the management of new business.

CONSTRUCTION AS A LOW-RISK BUSINESS. Being producers of custom-made products, contractors have been spared the need to sell after the products have been manufactured, unlike the makers of automobiles, home appliances, and other consumer goods. As a result, contractors have placed the highest priority on bringing in orders, and have made less serious efforts to raise production efficiency and trim production costs. Consumer goods makers set retail prices first and then try to garner profits by trimming costs as much as possible, but contractors have determined prices by adding the necessary costs onto a profit margin, whether the pricing is based on the lump-sum method, as in Japan, or on the cost-plus-fee method, as in Western countries. Thus, it has been relatively easy for the contractors to pass the risks onto the client and subcontractors, except for overseas projects and domestic development projects, making the construction contractor trade largely risk free. Nevertheless, this advantage may become a disadvantage in new business because of the contractor's inexperience in risk taking.

DEALING WITH GENERAL CONSUMERS. Contractors usually handle business contracts involving hundreds of millions of yen, but many of the new business ventures they start will involve products bearing price tags of tens of thousands of yen, thousands of yen, or even tens of yen. Put differently, products will have a greater added value and will be sold by the pound or ounce rather than by the ton. At the same time, the clients will change from the individual to the mass of the people—a brand-new experience for contractors other than home builders. The new business arena will demand different marketing, sales, advertising, and administrative techniques in order to sense customer needs and get consumers to loosen their purse strings by presenting the most irresistible products or services. In order to stay ahead of consumer needs, contractors will have to adjust their corporate organizations to new business styles.

The most difficult part of dealing with general consumers is their unpredictability: Nobody knows for sure what will sell best. Nothing is new about ice cream cones, but when one ice cream stand in Tokyo's sleek Aoyama District offered ice cream cones with a free choice of peanut, chocolate, or other kinds of diverse toppings, shoppers happily stood in a 30-minute waiting line on the sidewalk. Consumer reactions are more and more difficult to predict, as consumers seem to be shifting their criterion for product selection from "good or bad products" to "I like it or I don't like it. " Now that basic product qualities are fully ensured by the manufacturers, consumers can afford to use their often finicky likes and dislikes as a yardstick, forcing the manufacturers to revise their mass production policy into a small-lot, large-variation production policy. The masses have receded, and subgroups of consumers with different preferences have emerged as the elusive targets for business entrepreneurs.

Overcoming New Business Risks Stemming from Management Activity

PINPOINTING INTERNAL OBSTACLES. The greatest obstacles to developing new business lines lie within the company organization. The bigger the company organization, the more staunchly conservative the employees. They become dedicated to maintaining their company organization, resisting reforms, and hindering any moves to revitalize the organization. Following are likely causes of the in-house conservatism of a large company:

1. Employees tend to become self-satisfied and to rest on the laurels of past successes.
2. They are liable to discourage any colleagues who are considered rivals in the company organization.
3. The work attitude of evading risks, reforms, and mistakes often results in habitual complaints about co-workers.
4. Employees are reluctant to make a move until their rival associates make a move, after which they attempt to frustrate the rivals.
5. The company lacks the ability to acquire advanced technologies from outside, or depends too heavily on purchased technologies and lacks research and development abilities.
6. The company tends to be slow in responding to market and customer needs.
7. The company emphasizes short-term profit performance, failing to take notice of future business potential. For example, the accounting and finance department of the company is likely to place top priority on budgetary and profit management, often raising objections to investments for new business ventures.
8. The company is prone to develop conflicts of interest between departments, between the main office and branch offices, between office workers and site workers, and between clerical and engineering workers. For example, because a new business operation could result in competition with a client or a group of clients, the sales department may oppose such new business.

Of course, these inner obstacles should be removed, and the employees must be given a sense of crisis to achieve the needed growth.

EVOLVING AS A DEMAND CREATOR. In the past, salespeople from the construction company made regular visits to would-be clients, inquiring about their plans to erect buildings. But today, just making frequent calls on and getting acquainted with prospective customers is not enough to harvest orders. Planning has become a crucial factor, to suit the client's needs within the budgetary and time limits indicated by the client. Going a step further, the contractor should create a demand for a building by showing an attractive plan that meets the needs that the client could not properly express. The importance of generating demand is certain to increase for contractors in their attempts to foster new business operations.

JUMPING INTO A BIGGER COMPETITION ARENA. Although contractors used to compete within the borders of the construction industry, today the boundary line is gradually expanding as they branch out and outsiders move in. Nippon Steel, the world's largest steelmaker, has begun to build public housing,* thus emerging as a formidable rival in that field. Nippon Telegraph and Telephone, Japan's largest communications enterprise, was converted from a government organization into a private business in 1986, and the giant corporation's repair and maintenance department has now become a threatening competitor for the construction industry, changing from its previous status as a major client. Now NTT also looms as a client-turned-competitor for the old "NTT Family," consisting of communications equipment manufacturers and communications engineering companies. Japan Tobacco is another recently privatized corporation which is expanding into the leisure industry, utilizing its extensive network of offices and other facilities. The six newly privatized railway companies are expected to develop into multiservice companies, including leisure, hotel, and travel agent services, on the strength of their passenger and cargo transport operations, and are already active in the development business using the vast properties they own around stations and alongside the railroad tracks, much to the consternation of construction companies.

STARTING SMALL TO GROW BIG. Since new business ventures require contractors to make speedier, timelier, more flexible decisions, the traditional bottom-up-and-everybody-participate type of decision-making process will be too time-consuming to respond to opportunities. New business naturally contains many unknown elements, and inexperience must be supplemented by the sometimes bold decisions of executives with a strong will to succeed in the new business.

*Public housing comprises mainly condominium apartment houses built by prefectural governments with financial assistance from the national government. Contractors have a strong interest in the growing demand for the rebuilding of older apartment units built during the 1950s and 1960s.

Inexperience can also be covered up by the orthodox method of learning from history and case studies of other companies. Ultimately, contractors will feel it necessary to find a universal law of growth that transcends periods, industries, and corporate characteristics.

The daily routine management techniques cannot be applied to new businesses which are fraught with unpredictable and unknowable factors. The best approach may be the feasibility study technique, whereby the next step is determined after completing and reviewing each stage. Contractors cannot move into only the greenest fields with the lowest fences, as before. They may have to climb into business fields with a high fence. The initial obstacles should be overcome with patience if the growth potential seems large, and if small successes are accumulated at the beginning, they will stimulate the company to seek innovations and reforms within its organization.

EMPLOYING PEOPLE WITH AN ENTREPRENEURIAL SPIRIT. A business is no more and no less than the people who run it. Since the construction company regards new business as a cornerstone of its growth strategy, the company should deploy its most talented staff for the new business, rather than to use new business as pastures for older, less productive employees kept on the minimum pay basis just to honor the lifetime employment arrangement. New business needs the all-out support of the company.

According to a survey of the long-term labor market by the Economic Planning Agency of the government, the large number of workers born in the postwar Baby Boom* period will be reaching the top management positions in the year 2000, but because of their large number, only one in four is expected to get a managerial position, whereas until recently three out of four were able to become managers, because of the increased number of positions created during the period of high economic growth and the seniority-based promotion system adopted by most Japanese corporations. But the mental hardships for the three-in-four failures would be less severe than those for the one-in-four failures because in the first case, the dropouts will be the majority rather than the exception. If these less motivated workers with fewer responsibilities form the main part of the company work force, the company has a bleak future. In this regard, new business will serve to whip up motivation among otherwise insensate employees. Contractors badly need workers who think as entrepreneurs and innovators in order to develop new business. Their future success hinges basically on the number of employees with entrepreneurial thinking.

*The Baby Boom commonly refers to those born between 1947 and 1949. Their number is markedly more than the Japanese born in other years, and they have been forced to experience extra competition in their educational and business careers. Now, companies are deeply concerned because their salaries are increasing while managerial positions are not.

PROMOTING INTEGRATION AND DECENTRALIZATION IN THE ORGANIZATION.* According to some U.S. organization experts, organizations pass through the following developmental stages:

1. Growth through creativity
2. Growth through leadership
3. Growth through a transfer of power
4. Growth through adjustment
5. Growth through cooperation

The United States is reportedly in the fifth stage, and attempts are being made to overcome the bureaucratic and rigid forces in corporations. One corrective move is to introduce matrix organizations. Another is to promote integration, such as the improvement of administrative work in the company, to better cope with changes in the business environment. Also, efforts are being made to encourage decentralization by transferring some of the decision-making power to subsidiaries and project teams, for example. A third approach is to place greater priority on profit performances and on office automation, in a bid to reduce indirect costs. The integration and decentralization approaches are designed to provide improved abilities to adapt to the environment, and are particularly important for Japanese contractors in their bid to adjust their organizations to the changing construction market.

Today, outsiders may see two (construction and civil engineering departments) or more separate companies operating under the same name of a general contractor, and insiders too have a similar impression. The staff at the research laboratory and designing department may feel that they are working in different companies. Today it is increasingly difficult to manage all these departments under a single line of command, and the situation will probably worsen in the future, due to the addition of new business ventures. Consequently, contractors will need to set up more subsidiaries and group companies and to shape the main office organization into a "slim-bodied" office to oversee the activities of the specialized group companies.

ACCELERATE COLLABORATIONS WITH NONCONSTRUCTION COMPANIES. Collaborations and partnerships with nonconstruction firms in Japan and abroad will become a vital ingredient for the development of new business by contractors. Such moves are already noticeable, including the interindustry

*The issue of integration versus decentralization in corporate organization stems partly from the prohibition of holding companies in Japan, which most often tempts Japanese companies to grow bigger, but not smaller. Any leading general contractor, for example, employs over 10,000 personnel. Although a large organization can yield a large total power, the attainment of the versatility and flexibility of small organizations remains an ever-lasting challenge for big companies.

efforts to develop carbon fiber reinforced concrete, new transit systems, and construction CAD systems. Partnerships with foreign companies will become more frequent, and to enter the most attractive partnerships, contractors must have something attractive to offer in a give-and-take relationship.

Along with the predicted increase in collaborations and joint ventures, small-scale partnerships through partial equity participation and the transfer of a number of officers and staff workers by contractors also will increase in Japan and abroad. Subsequently, there should be an increase in corporate mergers and acquisitions by contractors. Since leading general contractors each have reserve funds amounting to billions of dollars, they can theoretically afford to buy other companies.

A FRESH IMAGE OF THE CONSTRUCTION INDUSTRY

The construction industry makes a large contribution to the cultural development of Japan through the construction of structures for living and for productive activities, and construction will continue to be an important activity in the future. Still, new business must be found by contractors, even though, in most cases, new business is difficult to find. The seeds of new business may lie in other industries, in overseas markets, in university labs, in customer needs, or within arm's reach of the contractor. These seeds are everywhere but are often too small to notice. Once discovered and planted, however, they will grow and bear fruit. The seed of new business will grow as an initially specialized market supported by a special segment of consumers is gradually developed by winning the acceptance of general consumers.

Construction is essentially an assembly work and system-making industry involving the combination of diverse technologies, and it provides both hardware and software services. As assemblers of knowledge and skills, contractors are highly experienced project managers and should be able to find and nurture new business seeds by taking advantage of their superb information-gathering and management abilities. If the buildings the contractors construct for their clients constitute the skeletal structure of society, then they must expand their operations to bolster the muscles and the central nervous system of society by going into new business ventures. Yet contractors will always build their new business ventures on the resources available from the main construction business, and a corporate group will be developed around each construction company through mergers, corporate acquisition, and establishment of subsidiaries. As stated in the concept of holons advocated by some economists, the power of the combined activities of each subsidiary or affiliate will be greater than the sum of the power of the group members. This is the ideal image of a future construction company and its group.

Tokyo Bay Hilton International (Japan) Shimizu Corporation

CHAPTER 5

AN INTEGRATED ENGINEERING CONSTRUCTOR STRATEGY

CONCEPTS OF AN ENGINEERING CONSTRUCTOR

ABC of EC

In 1971 Taisei, a leading Zenecon, established an engineering department in its main office technical headquarters, to handle the piping and mechanical installation work linked with industrial plant construction projects. Although other general contractors had had similar ideas, Taisei was the first construction company to form an engineering section in Japan, and this was observed with a great deal of attention by the managers of other companies. Subsequently the oil crisis occurred, and Japanese contractors began to seriously contemplate engineering as an undertaking for coping with the beginning of a slow economic period.

EC stands for "Engineer Constructor," a term introduced by the Bechtel Group of the United States in its company guidebook, "The Bechtel Story." Today, Shimizu Corporation uses these initials to represent "Engineering Constructor," while Kajima uses them to mean "Engineer Constructor," as defined by Bechtel. Takenaka uses "Engineering Contractor," and Taisei uses it for "Engineering and Construction." These subtle variations in terminology indicate the slight variations in Japanese contractors' ideas of EC, but their approaches to the task of developing their own engineering capabilities are basically similar.

Construction companies have acquired a wide variety of hardware and software technologies through the construction of all types of structures. In addition, contractors have always operated in an assembly industry, which produces products by combining various technologies, and while engaged in assembly work, the contractors have achieved an excellent capability for organization. For these reasons, contractors have a greater potential to develop an all-around engineering operation, more so than specialized engineering companies and industrial plant manufacturers who operate only in a specialized field. Contractors believe that they can move into the engineering market by combining planning, design, and other diverse abilities with their main construction operation.

EC Functions

The functions of the Engineering Constructor are more extensive than those of the contractor (see Figure 5.1). As indicated by stage [1] in the figure, the EC carries out a number of important activities in the planning stage, including the tapping of project potential, environmental assessment, feasibility studies, and consultation with the prospective client to clear all problems that may block the realization of the planned project. In the design stage, [2], the EC provides the basic and detailed design of the building structures and mechanical installations. In the procurement and construction work stage, [3], the EC not only builds housing structures, but also

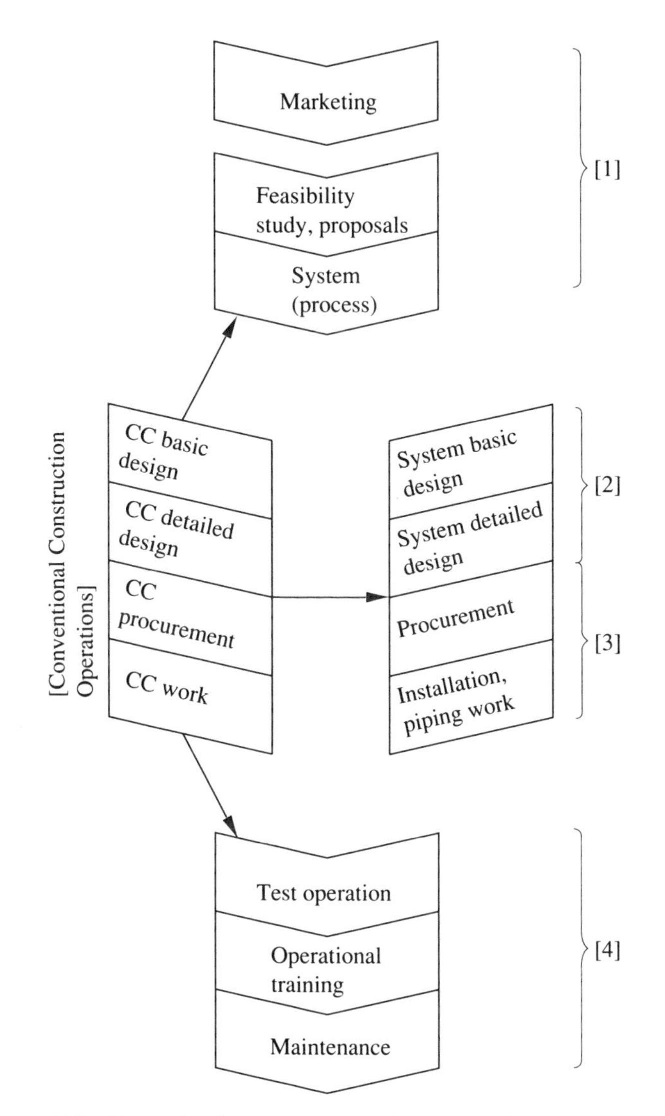

Figure 5.1 Expansion from a contractor into an engineering constructor

procures and installs all the necessary electrical, mechanical, and piping hardware systems. In the final stage, [4], after test runs, the EC delivers the product with a full turn-key service consisting of operator training, instruction, and maintenance.

As construction companies gradually expand into the machinery engineering and system engineering field, the engineering term "facilities" has come to be used interchangeably with the construction term "buildings," suggesting that contractors are steadily putting down roots in the new engineering field. Today, the contractors even apply EC expertise to urban redevelopment projects because of the wider range of services involved in those projects, from demand-creation type marketing and feasibility studies to maintenance.

GROWING EXPECTATIONS FOR ENGINEERING

Looking for Full Turn-Key Orders

Previously, when clients wanted new plants, they invariably hired contractors for housing structures and machinery makers for equipment. But in 1982, Takenaka Corporation erased this common usage by obtaining a full turn-key order for a fully automated FMS (flexible manufacturing system) plant from a steel product maker. For a contractor to win full turn-key orders, the company must take part in construction projects from the planning stage and must demonstrate its software capability. The plant department staff of Takenaka Corporation began with the study of the FMS, designed the production system, determined the plant floor layout, selected major machinery and equipment, and drew up the basic plan for the plant housing structure, side by side with the client. In this planning process, Takenaka was able to win the confidence of the client and thus received the full turn-key order.

In the same year, Shimizu Corporation, with the cooperation of Mitsubishi Corporation, a large general trading house, received a full turn-key order for the construction of a Solar Energy Research Institute from the Iraqi Agency of Science and Technology. The contract included, in addition to the procurement and installation of the required furnishings and equipment, maintenance and local engineer training services for one year after the completion. Since Takenaka and Shimizu won these contracts, a number of full turn-key contracts have been registered by Japanese contractors both in Japan and abroad, particularly from foreign high-tech companies wishing to build plants and laboratories in Japan. Foreign companies tend to ask Japanese contractors to supply a wide variety of services, ranging from selection and acquisition of the plant site to coordination with local residents, selection of machinery makers, and recruitment of plant workers. In the past year, this type of order included contracts signed with Motorola, Intel, Warner-Lambert and IBM of the United States and Novo Industry of Denmark. Although some of those projects are not completely "full" turn-key orders in the true engineering sense, they will help contractors to further advance their engineering abilities.

Developing New Markets through New Technologies

Large Japanese contractors are aggressively pursuing the development and accumulation of new technologies related to construction, and the products that have resulted from such technologies include high-tech clean rooms, air-supported membrane structures, coal storage concrete silos, and ocean ranches (see Table 5.1).

HIGH-TECH CLEAN ROOMS. Fueled by increased semiconductor production over the past years, the high-tech clean room market has expanded more than 20% every year to reach an estimated 300 billion yen ($2 billion) in sales per year. The market requirement is for high-tech clean rooms with a high purification capability, and the market is steadily expanding from the 80% share held by clean rooms for semiconductor production to those for surgery and sterilized rooms in hospitals, for animal experiment rooms, and for industrial clean rooms in drug and food processing plants.

Accordingly, leading contractors have been busily involved in experiments using prototype high-tech clean rooms to continue their technological advance in this field. Shimuzu and Kajima both have model clean rooms for electronics and for biotechnology use. In these prototype models, air purity levels, earthquake resistance, energy efficiency, and other performances are measured to obtain reliable data and to develop integrated clean room system technologies. On the strength of its proven technologies, Kajima doubled the orders received for high-tech clean rooms from 7.5 billion yen ($50 million) in 1983 to 15 billion yen ($100 million) in 1984 to 30 billion yen ($200 million) in 1985.

Table 5.1 Prototype Models of New Engineering Facilities Built by Japanese Contractors

	Shimizu	Kajima	Taisei	Takenaka	Ohbayashi
High tech clean rooms	1983	1982	1984	1985	1983
Coal storage silos	1984	1982		1984	1981
Air supported membrane structures	1983	1985		1982	1983

AIR-SUPPORTED MEMBRANE STRUCTURES. Japan's first roofed baseball stadium opened in Tokyo in the spring of 1988, heralding the coming of air-supported membrane structures (ASMS). This was 10 years behind the first ASMS baseball park built in the United States, largely because the Japanese construction standards law did not allow, until recently, the use of chemical fiber membranes as part of permanent building structures. Shimizu, Ohbayashi, and Takenaka (the contractor for the first Japanese roofed baseball stadium), run prototype ASMSs to advance the performance of the room membranes, membrane inflation systems, antiwind stability, safety, and indoor comfort (lighting, air circulation, temperature, humidity, acoustic effects, etc.) functions.

As a result, a number of ASMS projects have been approved by the Japan Construction Center, a public authority, including a gymnasium built by Takenaka Corporation, an indoor school swimming pool by Taisei, an indoor tennis club by Kajima, and an event hall in a zoo by Shimizu. These completed structures are medium sized, but contractors are certain that a large new market will form in the future as the demand for larger ASMS increases to house large sports facilities, multipurpose event halls, supermarkets, and warehouses among others.

COAL STORAGE CONCRETE SILOS. Since industrialists have renewed their interest in coal as an economical source of energy after two oil crises, contractors have stepped up their efforts to promote coal silos. Through joint development activities with machinery makers and the operation of prototype silos to establish practical silo technologies, the contractors have commercialized new coal silos made of concrete materials. Shimizu Corporation for example, developed, in collaboration with Mitsubishi Heavy Industries, a "Wedge-Shaped Hopper Coal Silo," which has been built in two paper mills. The "Big Atlas Silo," developed by Kajima hand in hand with Hitachi Metals, is now in operation at an oil refinery, and two units of the "W-Conical Hopper Coal Silo," completed by Ohbayashi with Hitachi are serving a power station and a synthetic fiber plant.

OCEAN RANCHES. In 1976 the United States and the Soviet Union both proclaimed an exclusive fishery zone extending 200 nautical miles from the shore line, in accordance with the revisions made in the International Sea Law in the United Nations. Japan, the world's largest fishing country, was hit hardest, and Japanese fishermen turned to coastal fishing, particularly the idea of raising fish in ocean "ranches." In 1984, the first ocean ranch was built in the Saeki Bay, Oita Prefecture. Unlike the conventional fish farms, the fish ranch does not have nets to keep the fish from wandering away, but keeps them within a certain area by conditioning them to certain sound signals that indicate feeding times. Shimizu Corporation supervises the engineering of this ranch, including the maintenance of an ocean solar power generation system, which Shimizu supplied to obtain the electricity required by the fish ranch system.

Contractors are also involved in the development of artificial undersea "rocky" structures serving as gathering places for fish. Hazama-Gumi is working on artificial mountain ridges designed to create new fishing areas; Penta-Ocean Construction is testing techniques to produce upsurge currents using artificial structures to attract fishes; and Ohbayashi is pushing ahead with the development of undersea structures particularly suited for use in shallow waters.

OTHER DEVELOPMENTS. In addition to the above-mentioned technologies, Japanese contractors are engaged in the research and development of concrete offshore oil platforms, underground liquefied natural gas tanks and new passenger transit systems, among many other projects. The contractors are hopeful that these technologies will eventually help them to expand the construction market.

ENGINEERING TECHNOLOGY AS MERCHANDISE

Engineering skills can be sold independently from construction work, and are classifiable into three types of engineering merchandise.

NONCONSTRUCTION ENGINEERING MERCHANDISE. This merchandise group contains water treatment systems, air-cleaning systems, and building control systems, all of which can be installed in existing buildings. Air-cleaning systems developed by Shimizu Corporation have already sold 260 units, now used primarily in buildings constructed by other contractors.

ENGINEERING SOFTWARE MERCHANDISE. Engineering merchandise does not always need to involve machinery. It can be engineering knowledge or know-how; for example, site assessment for the planned construction of a supermarket or a factory, and computation of the future balance sheet for a research and development project. Also, demands for environmental assessment are increasing. These engineering software activities have increased as a natural consequence of contractors' taking part in a greater percentage of construction projects from the planning stage. In addition, orders for design are now on the increase both in Japan and abroad.

OTHER SOFTWARE MERCHANDISE. The most popular software merchandise is the computer program. There is a growing movement by large general contractors to sell to their smaller counterparts computer programs for design CAD, planning CAD, on-site construction project management, and structural computation.

ENGINEERING ACTIVITIES OF THE BIG 5

The engineering activities of the Japanese Big 5 general contractors are summarized below.

SHIMIZU CORPORATION. While continuing to operate as an all-around contractor in both the construction and civil engineering fields, Shimizu is particularly active in bidding to develop new markets. The company is engaged in a wide variety of new technologies not directly related to construction, including fully automated factories; air-support membrane structures; coal storage concrete silos; community air conditioning systems for urban redevelopment projects; chilled rockbed storage systems for urban redevelopment projects; chilled rockbed storage systems, which use old mines to store fruit and vegetables by tapping geothermal energy from Japan's abundant volcanic activity; concrete offshore oil platforms; and ocean ranches. The company's engineering headquarters has received numerous full turn-key orders in collaboration with other departments. Shimizu provides two types of employee training courses designed for EC operations: one on the execution of engineering projects and the other on the training of project managers.

KAJIMA CORPORATION. Compared to other Big 5 members, Kajima is more interested in the bolstering of EC abilities in civil engineering, and is particularly active in projects involving nuclear power stations, underground tanks for liquefied natural gas, coal silos, oil tanks, and pumped storage hydroelectric plants. Kajima has successfully carried out various joint research studies in construction, energy, and development activities with companies in nonconstruction industries, and in 1986 alone the company announced eight different technologies developed hand-in-hand with nonconstruction partners, such as easy office partitioning panels developed with NEC Corporation and a solar water heating system. Since the establishment of an in-house "EC business promotion committee" in 1981, Kajima has forged ahead with its companywide EC project promotion, strengthening interdepartmental cooperation under the leadership of the construction and civil engineering departments.

TAISEI CORPORATION. Taisei has built many production, distribution, and research facilities on a full turn-key basis. The company's engineering operation features an active use of outside technologies, and its particularly close relationship with machinery makers has enabled Taisei to receive many full turn-key orders. The company not only promotes companywide EC movements through its engineering headquarters but also tries, through the engineering marketing department in its international operations headquarters, to expand overseas engineering operations, particularly agricultural land development, hospital construction, food processing plants, and other projects for developing countries.

TAKENAKA CORPORATION. Takenaka emphasizes engineering operations in the fields of energy, high-tech industries, and social and regional development. The company is also noted for its receipt of many turn-key orders for high-tech plants from foreign companies operating in Japan. Takenaka promotes its EC strategy through 12 different project headquarters including the plant, environment, special structures, and development-planning headquarters. The medicine-welfare headquarters, for example, commands a total consultant capability in hospital construction projects, utilizing know-how of hospital business planning and administration. In 1984 the company developed a "visiting area analysis" computer program to enable the prediction of future patient turnouts for any hospital. This program improved planning and marketing abilities for the promotion of hospital projects. Takenaka is bolstering its computer-aided planning ability to promote orders by participating in projects from the planning stage.

OHBAYASHI CORPORATION. Although Ohbayashi seems to lag behind the other Big 5 members at the EC level, the company has handled a number of important engineering projects and is specialized in certain strong engineering areas such as concrete silos, "fine buildings" (buildings capable of providing precisely controlled environments), automated high-rise warehouses, and underground dams. In addition to coal storage concrete silos, Ohbayashi has mastered superior technologies in grain storage concrete silos through the acquisition of a related technical license. The company promotes EC strategies through an engineering operations department under the supervision of the in-house "engineering operations committee" chaired by the company president.

AN OVERVIEW OF NEW ENGINEERING ACTIVITIES

As a result of their continuing efforts to evolve into engineering constructors, the leading Japanese general contractors have been able to register an increase in full turn-key orders, new business orders, and orders for engineering technologies. But it is still too early for optimism, because the number of these orders is still too small to make any significant contribution to the contractors' total sales or profit performance. All EC strategies are meaningless unless they provide a significant increase in the company's profits. To boost profits through engineering operations, the contractors will have to raise their project completion efficiency and handle a maximum number of projects involving engineering technologies and know-how. In addition, the contractors must transform themselves from their current status of construction-centered engineering companies into truly integrated engineering companies with a balanced strength in nonconstruction operations.

Today, the EC strategies of Japanese construction companies are reaching the backstretch, and the number of competitors is predicted to increase as some of the smaller contractors intend to take an EC approach to differentiate themselves from the rest of the pack. Furthermore, the competition with nonconstruction industries will increase as specialized prefabricated engineering companies, electric machinery makers, and heavy machinery makers move into the construction market. In view of these impending actions, the following three points may be important for contractors.

Playing a Principal Role in the High-Tech Revolution

While the effects of high technology surge through the business community, construction companies have remained more like onlookers than surfriders. They have built housing structures for semiconductor production plants and for attached high-tech clean rooms, but have not yet won full turn-key orders for semiconductor plants inclusive of production equipment. The same is true for nuclear power plants, where Hitachi, Toshiba, Mitsubishi Heavy Industries, and other heavy machinery makers are assigned the main part of the contract (i.e., the power generation installation) while construction companies are asked merely to build the containers. In fact, contractors have served as sideline engineering companies in almost all of the construction projects in the high-tech field, able only to watch the machine makers play the main role because of their strong engineering expertise.

It is high time for contractors to put an end to playing a supporting role. They should strive to become engineering leaders in high-tech–related construction projects: more specifically, to become the prime contractor in semiconductor plant, nuclear power station or similar high-tech construction projects, thus obtaining all of the engineering work for the facility project. This can be done only when the contractors have reached a technological and engineering level higher than those of machinery makers.

Some of the Zenecons have already made a move toward becoming engineering leaders. Takenaka Corporation, in cooperation with a conveying equipment maker, has developed *Clean Shooter*, an automated conveyance system used in semiconductor production, which adopts a floating method to achieve a dust-free, vibration-free, fully automated, and maintenance-free conveyance system. Takenaka is eagerly promoting the Clean Shooter in the semiconductor industry, and is expected to use this new technology as a springboard to obtaining full turn-key orders from semiconductor companies in the future. Fortunately, the high-tech revolution is spreading from electronics to new materials, biotechnology, and marine and space development. There is a good chance that contractors will emerge as engineering leaders in these advanced fields.

Responding to the Need for Social Development Projects

The second point in future EC strategies has to do with the intensifying problems of the population explosion and food shortages in developing countries; the aging of populations and the resulting adverse side effects in industrialized countries; and environmental pollution and the natural resource and energy shortages on a global scale. In Japan, additional problems are emerging as the nation's industrial structure and the structures of cities and other regional communities undergo a major change that is due to the advance of new technologies, information and communications capacities, population aging, and internationalization. In attempting to deal with these problems, Japanese people show a greater desire to develop new social systems and implement corresponding construction projects on a long-term basis.

Although common definitions do not exist, *social development projects* may be divided into three different types.

1. Improvement of the living environment, including housing, medicine, health, education, and leisure aspects of everyday life.
2. Development and improvement of national infrastructures, including regional communities, cities, traffic and transportation structures, and information and communications networks.
3. Improvement of industrial infrastructures, including plant relocations, waste treatment facilities and commodity distribution facilities.

Of these agendas, urban redevelopment is in the spotlight today. In Tokyo, the Akasaka-Roppongi District Redevelopment Project is now completed with construction costs of 80 billion yen ($530 million; see Table 5.2). The 180 billion yen ($1.2 billion) Sumida River District Improvement Project and the 15 billion ($100 million) Shinjuku Nishitoyama Civil Employees' Housing Rehabilitation Project have just been launched in Tokyo. A large number of urban redevelopment projects in many different cities are underway or on the drawing board, often involving the modernization of downtown blocks around railway stations by individual landowners or by groups of landowners. The Nomura Research Institute has come up with an optimistic projection of 21.5 trillion ($140 billion) total spending to be generated by urban redevelopment projects in Japan in the next 10 years, and contractors are naturally excited by this pleasing outlook.

Entering headlong into the era of an aging population, private homes, hotels, and public homes for the elderly are springing up throughout Japan. Today, the priority is placed on increasing the number of these facilities to meet the growing demand,

Table 5.2 Major Urban Redevelopment Projects Underway

Project	Outline	Cost	Entrepreneur
Akasaka-Roppongi District Redevelopment Project	Construction of six buildings, including offices, hotels, condominiums, on 5.6 hectare site	$530 million	Mori Building Company
Sumida River District Improvement Plan	Construction of 40-story buildings consisting of 2500 condominium units on 9 hectare site	$1.2 billion	Housing & Urban Development Corporation, Tokyo Metropolitan Housing Corporation, etc.
Shinjuku-Nishitoyama Civil Employees' Housing Rehabilitation Project	Construction of three 25-story buildings condominium units	$100 million	Shinjuku-Nishitoyama Development Company

but soon the emphasis will be shifted to supplying high-quality facilities to provide elderly people with truly comfortable living conditions. In this connection, contractors are particularly interested in the ongoing construction of the Retirement Community in the hot spring resort of Itoh City in the Izu Peninsula by Misawa Home, a leading specialized home builder. In this community, Misawa is building permanent household-type country homes for senior citizens, with tomato hydroponic greenhouses to combine a healthy hobby with a steady income. This project suggests that contractors must apply their EC abilities to the development of new engineering systems desired by local communities, in order to be successful in future social development projects.

Commercializing Engineering Know-How

The third point in EC strategies pertains to the sale of engineering know-how. Although contractors have used their original technologies and know-how exclusively for their own construction projects to differentiate themselves from competitors, the time is ripe for EC-conscious contractors to alter this policy. Recently, demand has increased in developing countries for the purchase of construction technologies and management know-how from Japanese contractors. In Japan also, design firms and smaller contractors are buying technologies and know-how from large contractors in increasing numbers.

Despite these developments, most contractors sell their technologies on a piecemeal basis, without a companywide merchandising and marketing policy, and today contractors must actively promote their engineering technologies and know-how in the general market as full-fledged commercial products. Following this line of thinking, Taisei has already established a subsidiary to sell information services for construction-related technologies. The company is also contemplating the start of a temporary manpower agency business, which will dispatch construction managers and assistant managers to the construction sites of other contractors to help with the management of their construction project from planning to completion. Shimizu Corporation in May 1985 set up a subsidiary, Technet, to sell Shimizu's "Building Diagnosis" technologies as its principal merchandise. Still, these commercialized technologies are few compared to the total number of skills the contractors have accumulated, and there is a growing need for the contractors to commercialize a much greater number of technologies in a systematic manner, including the arrangement of technologies into merchandise packages.

STRENGTHENING FOUR ENGINEERING FUNCTIONS

Many more challenges must be met before Japanese contractors can call themselves engineering constructors. Particularly large problems loom in the financial and personnel areas. Furthermore, many of their technologies must be improved, or new technologies may have to be acquired through research and development or purchasing. At a minimum, contractors should bolster the following four functions.

MARKETING FUNCTIONS. For EC strategies to succeed, constructors must find projects that will utilize their technologies and hardware products. In the vital marketing process, social and market needs must be accurately determined through detailed market research and the necessary technologies and products developed. Then active sales promotions must be launched using a wide variety of advertising and publicity media. The key is to plan products ahead of the trend in social needs so that the marketing serves to waken sleeping needs.

TRAINING OF PROJECT MANAGERS. Project managers working for construction companies have been involved only in the management of construction work. Even if a project order has been obtained inclusive of both design and construction work, two different managers have usually been employed to oversee the design and the construction, respectively. But in the case of EC projects encompassing the design, construction, and after-completion maintenance stages, a single manager must control the project costs, time schedules, work quality, and so on throughout all stages of the project. From the standpoint of the client also, it is convenient and advantageous to settle questions arising from various stages of the project by communicating with a single responsible person. Contractors must therefore foster a sufficient number of project managers capable of controlling all stages of an EC project in a systematic manner.

INTEGRATION OF THE CAE SYSTEM. The CAD (Computer Aided Design) and CAE (Computer Aided Engineering) systems are useful in saving the human labor required in engineering work and in expanding access to engineering technologies from specific individuals to any employees of the company. CAD systems were initially applied to the computerization of design work, but today they also can be used to formulate project plans and provide presentation materials. In fact, the use of computers is spreading to the construction work stage, such as the planning of materials and labor procurement and the management of on-site work procedures.

Nevertheless, today's CAE systems are specialized in either planning-design work or construction work, and it is impossible to link the two stages. For example, if a change is made in the design, the computer does not make appropriate changes in the constructors work drawings and charts used on the construction site. Future CAE systems should be able to assist at all stages of each project (see Figure 5.2).

COOPERATION WITH NONCONSTRUCTORS. Contractors are already engaged in many cooperative activities with companies outside the construction

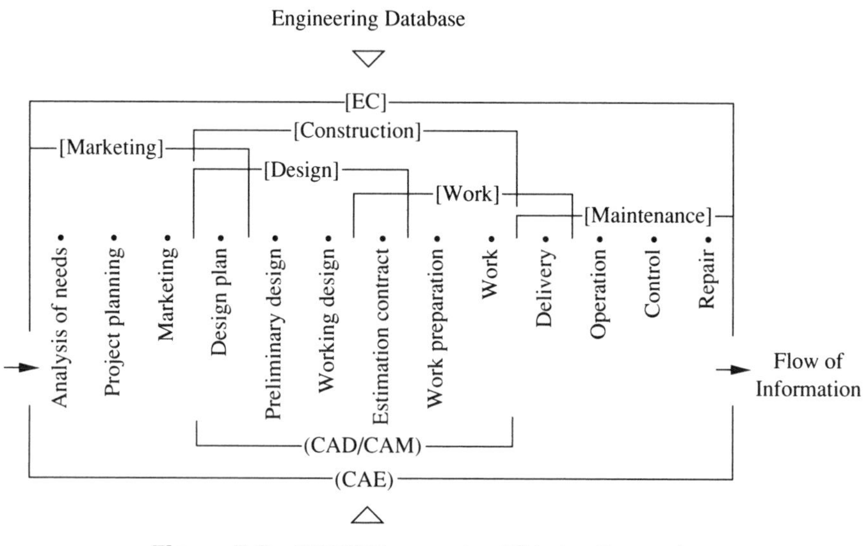

Figure 5.2 CAD/CAE concepts of Shimizu Corporation

industry, but the majority of these collaborations center around joint technological developments, exemplified by the development of coal storage concrete silos. Contractors hoping to evolve into engineering constructors need many more technologies and functions than they have at present, and must enter into not only joint technological development partnerships, but also sales tie-ups, joint ventures, consortia, and various other types of partnerships to function as a capable EC. Their engineering level can be raised by effectively utilizing the management resources of other partners from nonconstruction industries.

Sun Heights Arai-cho (Japan) Shimizu Corporation

CHAPTER 6

DEVELOPMENT PROJECT STRATEGY

DEVELOPMENT PROJECT OPPORTUNITIES FOR CONTRACTORS

Today, the Japanese construction industry is dancing to the music of development projects. Leading Zenecons want to eventually chalk up 10 to 30% of their total sales from development projects. The term *development projects* means construction projects involving sales and purchases of land or floor space and intended to promote regional and industrial development. Contractors hope to shore up their steadily declining profitability in construction by combining land and space, which are scarce and expensive commodities in Japanese metropolitan areas.

Contractors have taken due notice of Fujita Corporation, which has, in sharp contrast with most other general contractors, steadily increased its profit for several years. Thanks to development projects, including real estate business, Fujita posted a current profit of 15 billion yen ($97 million) for the fiscal year ended November 1986, thus exceeding its all-time high profit registered in the 1973 construction boom. To other general contractors, Fujita is proof that development projects can be lucrative.

Contractors are not the only group anxious to cultivate development projects. The real estate industry is the original developer and is naturally eager to expand its development business. In addition, the railway, steel, insurance, banking, and other industries are moving into the development field. Up to now, contractors in most cases have handled construction work for the development projects under contract from these construction companies, but in the future, the contractors must compete directly with these clients in the development business.

What are the factors that differentiate contractors from nonconstruction companies in the development business? The principal factor is the extensive role played by contractors in their clients' development projects as a representative negotiator with local residents, and as a designer and constructor. Contractors have a great deal of experience in planning and running development projects while satisfying the acreage, budgetary limits, and various other needs of the clients and tenants. This is the strong foundation on which contractors can build their development business. Development can include a wide variety of business ventures, from the construction and sales of condominiums, operation of rental buildings, and transaction of real estate products to exclusive or partial ownership of hotels and stores. In all cases, projects must be created, and the contractors' weak point lies in this creation of development projects.

Finding Land for Development

Development projects are possible only when land or buildings are available; contractors must find these pieces of merchandise to start their development business. A popular way of acquiring development lots is to purchase vacated plant sites from client manufacturers, but contractors are also trying new methods of acquiring land for development projects.

SETTING UP A LOCAL OPERATION BASE. The Fujita Corporation places a top priority on the acquisition of land to tap construction demand. The contractor not only hunts for large-acreage lots but also patiently looks for small plots. To locate would-be development lots, Fujita has deployed more than 50 *base offices*, staffed by three to four employees, in the Tokyo area alone. These base offices are dotted throughout downtown Tokyo and in suburban commuting towns, especially in areas around railway stations and in shopping neighborhoods. An important task of the base office is to find prospective plots for development projects. Then, if a number of landowners and lessees are involved, the base office staff attempts to persuade them to launch a construction project collectively, to adjust their differing interests, and to ensure the procurement of construction funds. If necessary, the base office staff spends a certain amount of money to buy plots to help bring a construction project into reality—an exceptionally large decision-making power to be entrusted by the main office to such a small outpost. Fujita apparently considers it necessary to bypass the typical, time-consuming bottom-up-and-everybody-participate process of decision making in order to succeed in the development business. By its actions, Fujita demonstrates the truth of the business motto, "Timing is what counts in the real estate business."

Both individuals and corporations have a persistent desire to own large buildings. But most fail to acquire them because of the difficulties of adjusting the conflicting interests of the landowners involved, of securing loans, of dealing with a heavy inheritance tax, and of attracting enough tenants among other difficulties. Fujita strives to solve these problems one by one, by establishing operations in local communities and winning the confidence of local residents. Thus, Fujita's development operation is strongly tinged with the characteristics of the real estate business.

ACQUIRING LAND AREAS RESERVED FOR PROJECT FINANCING. When private enterpreneurs carry out land consolidation in accordance with the law to provide sufficient roads, parks, and other public structures and to upgrade housing conditions, they usually finance the land consolidation projects with government incentives and proceeds from the sales of land areas reserved for this funding purpose. These reserved land areas are supplied by private landowners, each donating a fixed percentage of their landholding. In an arrangement dubbed the "Shin-Matsudo Method," Shimizu Corporation became the first contractor to acquire a reserved land area as a developer, in exchange for the use of its own funds to finance the land consolidation project that the contractor, together with Mitsubishi Estate (Japan's largest real estate company), had been carrying out in Matsudo City on the outskirts of Tokyo since the early 1970s. The reserved land area was "sold" to Shimizu on the condition that the contractor would develop high-rise and medium-rise buildings housing a total of 5000 condominium units for sale. With room layouts emphasizing living comforts, the majority of these condominiums were sold out on the day sales started, thus making the Shin-Matsudo land development method a success.

ACQUIRING FLOOR AREAS RESERVED FOR PROJECT FINANCING. The acquisition of a reserved land area is a strong attraction to contractors in land consolidation projects, and in urban redevelopment projects, the contractors may look forward to getting reserved floor areas in exchange for financing part of the project costs. As in the case of land consolidation projects, reserved floor areas are donated by the owners of the planned building to finance the project costs, and if some of the landowners wish to purchase the reserved floor areas, they can do so. As suggested by the now popular term Urban Renaissance, coined by former Prime Minister Nakasone to mean a revival of affluent community life through urban redevelopment, there is a growing demand for a rebuilding of old, substandard urban districts into modern commercial/residential areas. Both the government and the business community see such urban redevelopment projects as powerful stimulants to the sluggish economy.

In reality, however, urban redevelopment projects outside of big cities run into the difficulty of not being able to sell reserved floor areas. The projects thus become difficult to finance. One reason is that the floor space of the planned building is often expanded to the maximum floor area ratio (FAR) permitted by the city planning code, to fund the project as much as possible. The FAR is often raised from 100% to an ambitious 500% or even 800%, requiring many tenants to fill the added floor space. Another reason is that plans to bring in large would-be tenants, such as general merchandise stores, are often aborted by protests from small store owner groups in the neighborhood. As a result, in an increasing number of cases, landowners appoint a developer in exchange for a guarantee to bring tenants to the reserved floor area or with a contract to buy the reserved floor area.

The responses of contractors are complex. On one hand, sharing risks with landowners is a must for winning orders, but on the other, the risks can become dangerously large, especially if these urban redevelopment projects take as many as 10 years to reach construction completion. Consequently, unlike the purchase of reserved land areas in land consolidation projects, contractors usually decline to buy reserved floor areas in urban redevelopment projects but, as a compromise, become part owners of the project management companies formed by the landowners or persuade key tenants to buy the reserved floor areas.

RESPONDING TO DEVELOPER COMPETITION. To prompt the financial participation of private companies, public administration offices now hold an increasing number of proposal competitions among prospective developers. Developers were recently pitted against each other in a contest to submit attractive architectural, business, and financial plans for the Canadian Embassy and Ohmiya Industrial and Cultural Center projects, for example. These contests commonly feature

1. Large-scale development projects.
2. Formation of corporate consortia for the competition.
3. A construction company as a leader of such a consortium.

The larger the development projects are, the greater the risks involved. A key to success in such a massive project is to form a consortium consisting of all the types of companies needed to realize the project. Construction companies are naturally in the center of the group as main project planners, designers, constructors, and maintenance companies. Once selected as a developer, the contractor must engage in different types of operations: not only the sales of condominiums, but also the rental and operation of housing units and commercial facilities. For this reason, the developer business has proved too complex for small contractors to undertake and has remained the exclusive territory of relatively large contractors.

Specializing in the Development Business

While large contractors generally carry out all-around business operations, Hasegawa Komuten is particularly noted for its condominium development business. The company has consolidated its Hase Ko (HK) brand across Japan through the sales of housing units to city dwellers during the period of high economic growth overlapping a construction boom. The company is duly credited with the establishment of a brand name in the construction industry, where concepts of product brands have traditionally been weak and scarce. Hasegawa Komuten implements development projects in a flexible manner, not only acting as the project entrepreneur, but also utilizing subsidiaries and other group companies, depending on the nature of projects. Today, however, housing demand has reached maturity in terms of quantity, and competition is based on housing quality. Hasegawa Komuten now sees a limit in its housing specialization and is gradually branching out into the nonhousing field of office buildings, stores, hotels, and other commercial structures.

Commercializing Development Techniques

Real estate companies and trust banks are particularly active in the commercialization of development techniques. Towa Real Estate Development Company, for example, sells "space development" know-how to achieve higher utilization of the

Table 6.1 Major Overseas Development Projects Since 1985

Contractor	Area	Main Facility	Cost	Partners
Shimizu	Phoenix, Arizona	Offices	$53 million	Westco, Mitsui & Co. (USA)
	Peking	Condominiums	$46 million	Peking, Nissho-Iwai, Nomura Real Estate, etc.
Kajima	Dallas, Texas	Offices	$130 million	Mori Building La Forret Dallas
	Long Beach, California	Offices, hotel	$670 million	IDM Corporation
	Melbourne	Offices	$43 million	Costin
	Shanghai	Hotel, offices, condomiums, etc.	$175 million	John Portman & Associates, American International Group
Taisei	Long Beach, California	Offices, hotel	$210 million	Stanley Cohen, Marubeni Corporation
Kumagai Gumi	New York, New York	Condominiums, offices,	$190 million	Zeckendorf
	New York, New York	Hotel	$130 million	Zeckendorf, Sala, Goldman
	Seattle, Washington	Offices	$89 million	First City Development, Sixth & Columbia Associates
	Hawaii	Resorts	$4 billion	TSA International, Harvart K. Horita Realty, Belhouse & Joseph
	Melbourne	Offices	$550 million	Essington
	Peking	Offices, hotel, condominiums	$130 million	Peking Kyoko Hotel
Takenaka	Peking	Hotel	$90 million	Japan Air Lines, etc.
Tobishima	Santa Ana, California	Offices	$190 million	Shneider

Table 6.1 *(continued)*

Contractor	Area	Main Facility	Cost	Partners
Aoki	Orlando, Florida	Hotel	$375 million	Tishman Realty & Construction, Metropolitan Life Insurance
	Dallas, Texas	Apartments offices	$240 million	Tishman Realty & Construction, Wing Vineyard, Gill Savings Association,
	Hong Kong	New town	$140 million	Leighton (Asia) Limited
	Shanghai	Hotel	$68 million	Shanghai Travel Agency
Hasegawa Komuten	New York, New York	Condominiums	$160 million	W. L. Haines
	New York, New York	Condominiums	$100 million	Eastside Associates
	Hawaii	Hotel	$130 million	Landmark Hotel

available land and to minimize tax burdens by forming joint ventures between the landowners and the developer, through an exchange of a part of the land and a part of the building to be constructed by the developer. Mitsui Real Estate offers a similar "let's" exchange-of-equivalents system to facilitate the solving of conflicting interests among landowners and developers. Yasuda Trust & Banking promotes a "land up system," whereby the bank is entrusted by landowners to use their land to yield profits. Some contractors have begun to follow in these footsteps, including Tobishima Corporation, which sells an "eternal system" designed to provide consulting services to landowners on ways to most effectively utilize their land possessions.[9] Shimizu Corporation has developed a computer simulation program called "SCOPE" (Shimizu Comprehensive Planning Evaluation system) for the evaluation of efficient land use.

Future Directions for Contractors' Development Projects

There are an increasing number of urban development projects carried out by Japanese construction companies in the United States and elsewhere (see Table 6.1). Overseas development projects require even more locally rooted relations and risk dispersions than domestic projects, and more diverse approaches are demanded, including the establishment of subsidiaries, collaboration and joint ventures with local companies, and corporate acquisitions. Although the Japanese overseas development operation has just started, in the domestic market large gaps already exist between contractors. Some contractors are running a development business as extensively as the leading developers in the real estate industry, and on the other hand, others are only beginning to develop one or two condominium buildings at a time. Depending on the development strategies of each contractor, the gaps will likely widen on both the domestic and international fronts.

ENTRY INTO A GROWTH STAGE

As the environment surrounding the construction industry changes, new business opportunities are arising around four factors.

1. Relaxation of government regulations related to development projects.
2. Growth in consumer demand and increased demand for new plants in high-tech industries.
3. Approach of a period of aged populations.
4. Advance of internationalization in many spheres of life.

Relaxation of Government Regulations

To stimulate urban redevelopment by private efforts and resources, the Japanese Construction Ministry in 1983 did away with a number of restrictions on development projects in order to promote the supply of housing lots in city areas. At the same time, the ministry initiated a subsidy scheme to help finance private development projects considered to improve the level of community living conditions. These measures have resulted in an increased supply of housing lots and have made it advantageous for landowners to erect larger buildings in a concerted effort. Thus, some of the bottlenecks have been removed, although the major bottleneck—the invitation of large-scale stores as tenants—remains uncleared because the regulation on the opening of a large store in a community of small local stores is under the jurisdiction of the Ministry of International Trade and Industry. Contractors do, however, have high hopes of an interministerial endeavor to remove this large obstacle to development projects. Overall, contractors welcome the government moves to introduce private resources, not only to urban redevelopment projects but also to the construction of roads, bridges, airports, and other social infrastructures that have been exclusive undertakings of the government until recently.

The Advent of an Information Age

The future course of development projects will be deeply affected by new media that will far advance the present level of information processing and distribution capabilities. If the INS (Information Network System) of Nippon Telegraph and Telephone (NTT) is put to practical use, doing office work, school studies, and shopping at home will become common, and the geographical areas suitable for development projects will expand from the centers of big cities to other areas. In fact, the selection of development areas will drastically widen, and the most suitable locations for different types of projects will become available at lower costs.

The Ministry of International Trade and Industry, for example, envisions the construction of model New Media Communities in many regions of Japan to promote the nation's transformation into a high-level information society. The Ministry of Posts and Telecommunications is planning its own version of future communications-intensive communities, called Teletopia. In addition, the proliferation of new-media information networks will require contractors to supply buildings with high-level communications and information systems for the convenience of building users. Accordingly, contractors will need to upgrade their information hardware and software expertise.

Also, the development of new media is closely linked with an increase in data processing speed, and this will boost the efficiency of each development project for contractors. The information currently filed by small realtors in each community will become more easily accessible, and exchanges of such information will be prompted. The types of information available will be expanded to include those on industry trends, interest rate trends, city development potentials, urban planning, tenant activities, competitor activities, market demand trends, and citizens' opinions and awareness. These wells of information, plus specialized information from clients and correspondent banks, will be applied systematically to promoting development projects. Accordingly, the information systems of small realtors will have to be upgraded for modernization of their businesses, and the concerted attempt by Century 21 of the United States and C. Itoh, a giant trading house, to link small realtors across Japan with a computer network service is an important step forward to a new age of information in the construction industry.

At the same time, contractors anticipate that the high-tech industries will set up an increasing number of new plants. In the United States, "office parks" and "business parks" have been built to house R&D–oriented facilities in vast areas of open land. In Japan, the Ministry of International Trade and Industry is pushing ahead with a long-term plan to build high-tech "Technopolis" communities with populations of 40,000 to 50,000 near larger cities with populations of 200,000 to 300,000. A number of local governments are also engaged in the development of R&D–oriented communities. In Chiba Prefecture, adjacent to Tokyo, a group of large corporations led by Nippon Steel and Shimizu Corporation, with the cooperation of the Chiba Prefectural Government, is building the Makuhari Techno Garden as a central unit of the Makuhari New Business Center on reclaimed land on the shore of the Tokyo Bay. In the Techno Garden project, the corporate group is constructing a complex of six intelligent buildings to house R&D laboratories, sports centers, and commercial establishments for completion in 1989.

The Approach of an Era of Aging Populations

As the Japanese population as a whole ages because of an extended life expectancy and a declining birth rate, the weight of products and services catering to senior citizens increases in the total market. Because of a greater percentage of older people, there will be a growing emphasis on home life and housing quality. Houses will be expected to provide not only living comforts but also health care and a community environment for older people to enjoy leisure, recreational, and social activities. Interest will grow in houses built specially for the aged. Contractors should tap the emerging senior citizens' market by combining their development project and new business strategies.

The Advance of Internationalization

As the human interchange between Japan and the world continues to expand, the demand for houses designed for foreign nationals will increase. This demand will spread from Tokyo and a few other big cities to regional areas because the residences of foreign nationals are spreading throughout Japan. The demand for international facilities such as hotels, conference halls, and exhibition pavilions is also likely to grow. Some cities aim at becoming convention cities as part of their programs to revitalize regional economies. Local communities will probably opt to bolster their international facilities as an important city function, and contractors will be able to expand their business opportunities by closely catering to regional needs.

PROMISING AREAS FOR DEVELOPMENT PROJECTS

Promising areas for development project may be grouped into four types.

1. The first concerns national infrastructures, such as projects to build a trans–Tokyo Bay highway, a trans-strait bridge between Awaji Island and the mainland Honshu, and a New Kansai International Airport offshore from Osaka. These huge projects combine diverse technologies and feature the introduction of private

finances and an increased participation of foreign companies. These are public works projects with an increased initiative by private companies.

2. The second type is local city development, including new town construction and urban redevelopment. With relaxed government regulations and an Urban Renaissance policy backup, these projects are predicted to show a brisk growth, and in addition to city development, the development of resorts is likely to accelerate, favored by government policy. Projects in this category also feature combinations of diverse facilities to improve community environments.

3. This category consists of social service projects, such as the construction of cultural, informational, and welfare facilities in response to changing social needs. The aging of the population and the advance of women into the job market are two such changes. These projects require contractors to place a particular emphasis on maintenance services and to acquire professional software know-how on the exact needs of the aged, women, and other prospective customers.

4. Finally, industrial promotion is an area in which many development projects can be expected in the future. Included in this category are the Technopolis, Teletopia, and other plans to promote regional industries. Contractors will be required to work closely with local administrations and local business communities.

All but the second type would have been implemented as public construction works up to now, but in the future, construction companies will be able to become developers of these projects as a result of the government policy to utilize more private resources in public works. Contractors have the construction technologies and know-how to obtain national infrastructure projects, but will need to cooperate with other industries to gain software know-how in order to better handle the remaining types of industrial development projects, including the promotion of convention business. Local urban development projects have been carried out mostly by real estate companies, but opportunities for construction companies are increasing. Another area on which contractors should place special emphasis is that of housing development.

The Future of Housing Development

An in-house document of the European Community Commission in 1979 likened the small Japanese houses to "rabbit hutches," which is now a well-known, if not well-accepted description of Japanese housing conditions. Another phrase indicative of Japanese housing is "Eight percent of homes vacant;" this is in dramatic contrast to the vast housing shortages of the postwar period. Today a housing glut exists because the rabbit hutches have outnumbered human families, particularly through a population migration from villages to cities. Another indicator is "Housing construction units slipping to one million. " The all-time high of 1.64 million housing construction units was recorded in 1976. These indicators have led to the newest indicator: "Shift from quantity to quality. " Today, developers must supply more comfortable homes with a satisfactory environment and pay close attention to the individual needs of each customer by offering, for example, free planning services for room layouts.

One predictable problem concerning the growing emphasis on housing quality is the difficulty of rebuilding old collective housing units. To solve disputes among the owners of condominiums in the same building, the Law Concerning Sectioned Ownership of Buildings was amended in 1983 to enable rebuilding of the entire condominium building by the consent of 80% or more of the condominium owners, but in reality, rebulding is difficult unless all owners give consent. Judging from the condominium construction rush that started from the second half of the 1950s, the need for rebuilding will peak toward the end of this century, and if a large percentage of what by then will be very deteriorated condominium buildings are owned by senior citizens, the owners may not be able to afford such rebuilding. Given this outlook, contractors should make technological improvements in the renewal and redevelopment of collective housing, and perhaps go as far as to design new condominium buildings featuring easy renewal or rebuilding systems. The house being the core of human activities, contractors should continue to emphasize the importance of supplying high-quality housing.

THEORY AND PRACTICE OF DEVELOPMENT PROJECTS

When planning and implementing development projects, the contractor may consider three ingredients: (1) the software of the project, (2) the cultural attitude of the contractor, and (3) development techniques (see Figure 6.1).

Software

Contractors should accumulate the know-how of supplying houses, stores, factories, and other construction products matching the exact needs of customers. On the home-building front, the demand is growing for houses for old aged people. These houses must, of course, be designed on the basis of detailed knowledge of old people's living behavior. Also important is that houses be designed in which two or three generations of a family can live in comfort. Similarly, housing for foreign nationals in Japan should be spacious and should also reflect their customs, cultural interests, and religious practices. Homes equipped with new communications equipment will be useless unless contractors upgrade the software for selecting the types of information required by the home dwellers. Contractors should not pre-

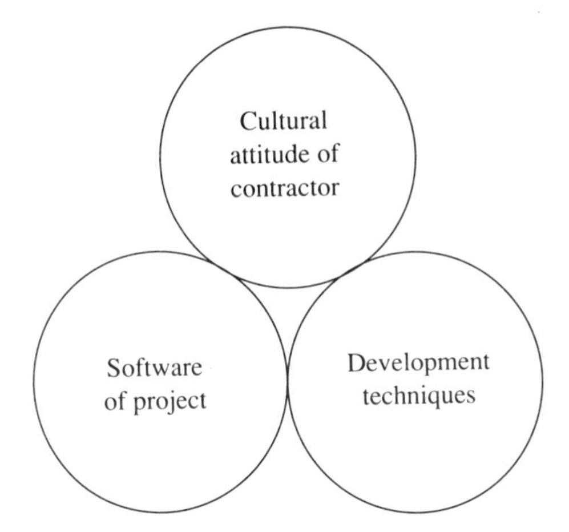

Figure 6.1 The three elements of development projects

determine whether their products will be built for sale or rent, since this should be determined on the basis of customers needs; this is particularly true for homes that may be sold or rented according to the life stages of prospective home dwellers. By trade, construction companies command more advanced software than that of other industries, and herein can be found their opportunities.

Another important point is a totalization of development projects. Contractors cannot simply develop housing units or office buildings but must also attract convenience facilities, such as stores, restaurants, sports centers, and cultural halls, to provide a more satisfactory community environment. Going a step further, contractors may start a business around the development of various facilities to improve community environments. Software, or know-how, of improving local community life will be required in future city development projects.

Company Attitude

To embark on development projects, contractors must systematically select regions with a large development potential. Then the contractors will be able to map out a clear future vision of the region and their involvement as project entrepreneurs. Because the people of the region expect the development project to contribute to a betterment of their community in all aspects, the contractor must relate with the host community on far more than just a business plane; the cultural plane is also extremely important. Producers of foods, cosmetics, apparel, and other consumer goods have for some time publicized the idea of cultural contact between corporations and consumers. Nevertheless, contractors, although they construct buildings and roads that are used by consumers on a daily basis, have never been in close contact with people on the street.

Contact with citizens is unavoidable in development projects. Citizens will judge whether the contractor of their development project has made a definite contribution to their community, and their evaluation is bound to influence the contractor's business opportunities in development projects of other regions. Culture being omnipresent, future contractors must not only build structures but also utilize their own corporate cultural resources to enrich the regional culture.

One type of involvement with a regional community may be the following process. As a contractor constructs a housing complex, block by block, the new residents of these blocks would form a neighborhood organization to set up a community promotion committee, of which the contractor would become a sponsor. The committee could start a contest for naming streets and neighborhoods and for designing symbols, and the contractor might present awards to the winners and also establish prizes for facilities contributing to the making of an attractive town, thus playing a leading role in the promotion of a community identity. The contractor should, of course, strive to shape this community's identity to suit the regional characteristics, taking special care to achieve a specific pattern of community development matching the unique features of the particular region. By doing so, the contractor wins the acceptance of the community residents on a cultural level.

Methods of Development

As the types of development projects have diversified, a greater variety of technical problems has arisen. The formulas of land trust, of exchange of equivalents, and of the new land leasehold* have been put into effect to solve some of the problems. These techniques have proved successful because they suit the needs of landowners, tenants, and developers. As developers, contractors must determine the needs of all the interested parties of development projects. They must also utilize existing techniques or formulate new techiques that take advantage of the recent relaxation of leasing and other laws. To this end, a partnership may be formed with trust banks or real estate companies to gain the necessary know-how.

In addition, it is often advantageous to form a consortium for executing a development project. Particularly beneficial is the participation of nonconstruction firms in such activities as fund procurement, the attraction of tenants and surrounding service facilities (e.g., restaurants and stores in the neighborhood), and conceptual designing of the project. The importance of a consortium grows in proportion to the size of the project, from the standpoint of risk sharing. Consortia can

*Under the new land leasehold formula, the leasehold is limited to one generation so as to lower the lease fee and to grant more power to the landowner than do conventional lease contracts.

also promote an interchange of talented workers among different industries, since contractors cannot supply all the necessary staff for development projects just by training their own employees. In fact, the formation of consortia is a must for construction companies wishing to capture a greater number of development projects. Nevertheless, consortia will be less beneficial if the contractor merely serves as a constructor, rather than a project manager in such consortia.

Kobe Marine Museum (Japan) Aoki Corporation

CHAPTER 7

TECHNOLOGY DEVELOPMENT STRATEGY

PROGRESS AND CHALLENGES IN TECHNOLOGY DEVELOPMENT

The current period is marked by phenomenal advances in technology. A rapid succession of new products has been introduced in fields such as electronics, biotechnology, new materials, and space technology. One can say without exaggeration, "They who control the technologies control the market." Accordingly, all industries are placing top priority on the development of new products, and the construction industry is no exception. It has become difficult to boost orders merely by building structures according to the design and time schedule demanded by the client, and only those construction companies capable of offering new technologies and services through research and development will continue to exist. Japanese construction technologies, which have reached a top level, are being diversified from the design and building applications into a wide range of peripheral areas, and the range of technologies explored by contractors will become even more extensive.

The extraordinary emphasis Japanese contractors place on technology and R&D is evident in their annual reports for 1985. Corporate statements such as these are common: "We are determined to ensure sustained growth of the company through the development of strategic technologies and the improvement of design and construction qualities"; "We intend to intensify our planning-and-proposition style marketing activities by improving our performance in technological development, engineering, and international operations, and to strive to make our construction business more competitive and profitable through improved productivity"; and "The company plans to step up its efforts to research and develop new technologies." Technological development strategies determine the allocation of research and development resources, are a key contributing factor for corporate growth, and are closely linked with engineering constructor, new business, and marketing strategies.

Surging R&D Investments

A construction trade newspaper company surveyed the sales and R&D investments registered by the 15 leading Japanese contractors in fiscal 1986 (see Figure 7.1). Shimizu Corporation led the research and development investments category with 11 billion yen ($74 million), or an amount equivalent to 1.1% of its annual sales. According to a yearly survey by the Management and Coordination Agency of the Japanese government, the construction industry as a whole laid out R&D investments amounting to an average 0.5% of total sales in fiscal 1986. Although not particularly impressive compared to other industries, this 0.5% is a high level in view of the fact that construction is an order-receiving business. The Big 6 in particular have boosted their research and development investments at an average

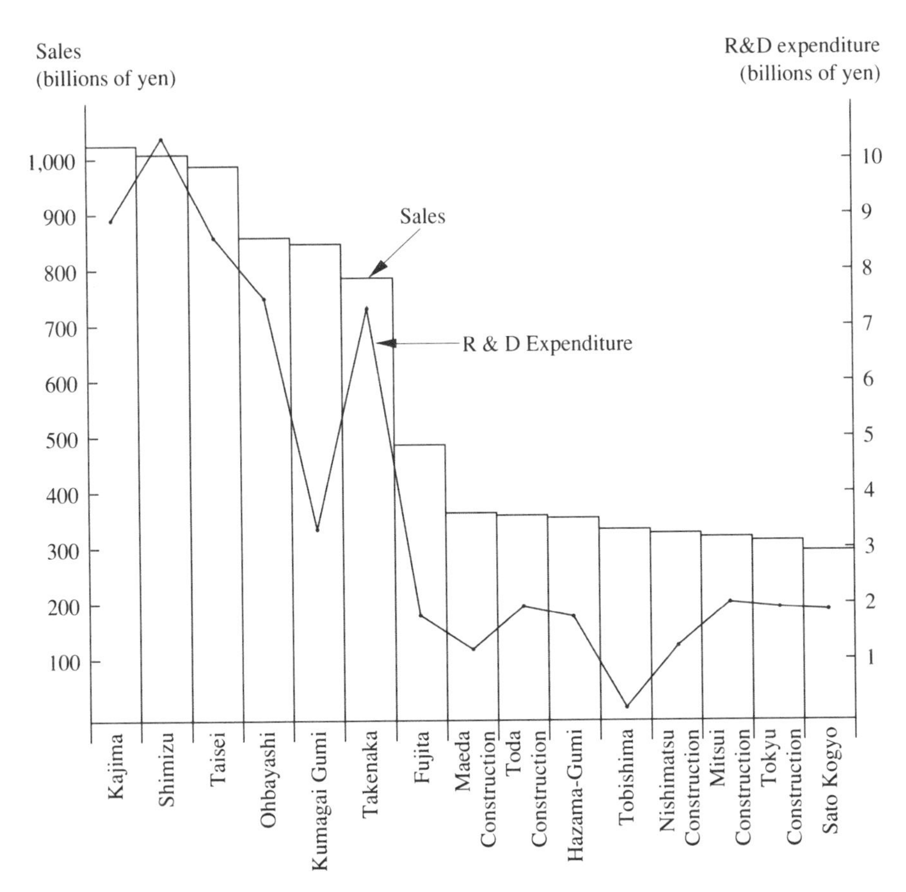

Source: Survey by *Nikkan Kensetsu Sangyo Shinbun*, a construction trade daily newspaper (Tokyo: June 23, 1986).

Figure 7.1 Sales and R&D expenditures of major contractors in fiscal 1986

rate of 9% a year for the past several years, and an increasing number of other major contractors have established or expanded their own R&D laboratories. Apparently all contractors are now aware that technological capability is a crucial ingredient for survival.

Background and Objectives of Technological Development

The Japanese construction industry boasts a long history of technological development. Shimizu for example, set up a design research section in 1944, which was expanded into a full-sized research department in 1950, and eventually into its current form as an advanced laboratory. The early researchers were engaged primarily in basic studies of construction materials and components, as the traditional wood-made shorings, scaffolds, forms, and other temporary items were rapidly replaced by steel substitutes. In the 1960s, research themes were expanded to include the mass production of housing units, with emphasis on prefabrication and work mechanization techniques.

Yet for large general contractors, the top-gear research and development period began much more recently, around 1980; their R&D investments sharply climbed from around 1980 (see Figure 7.2). Between 1980 and 1985, Shimizu multiplied its annual R&D costs 2.6 times, from 3.4 billion yen ($23 million) to 8.8 billion yen ($59 million), whereas its annual sales rose only 60% from 650 billion yen ($4.3 billion) to 1,050 billion yen ($7 billion) during the same five-year period.

One reason for this phenomenon is the heated competition caused by a market slump around 1980. To stay alive in the rough competition, contractors were eager to develop low-cost structures and cost-saving construction techniques. Another reason is the rise of electronics, precision machinery, and other high-tech industries since that time. As exemplified by high-tech clean rooms, the requirements of these advanced industries have prompted contractors to elevate their technical standards. Today, the contractors are continuing to acquire advanced technologies in all industrial fields and are attempting to combine these shiny new technologies with their construction know-how to supply improved construction services.

The third reason is the penetration of high technology into the social fabric. The widespread availability of high-level information and communications networks has led to a demand for high-tech features in construction (e.g., intelligent buildings). Since the manufacturers of information and communications equipment and other sophisticated machinery are moving in to meet this high-tech need in the construction market, contractors are compelled to master advanced technologies to compete with the incoming nonconstructors.

The fourth reason involves the saving of natural resources and energy, a never-ending theme in resource-poor Japan. Contractors are keen to develop energy-saving technologies to reduce the overhead costs of buildings for their clients. They also have a growing interest in water-saving systems, as big cities begin to develop water shortages, due largely to an excessive concentration of the population in metropolitan regions. One promising technology is a "building water recycling system" designed to recycle used drinking water as cleaning water within the building. Today, these technologies can be a powerful asset in pulling construction orders away from competitors.

The fifth reason for the contractors' heightened interest in technological development is a predicted diversification of the types of construction sites in the future. The New Kansai International Airport construction project is now underway on reclaimed ground offshore from Osaka, the first offshore airport in Japan. As a result, contractors have made special efforts to improve their reclamation tech-

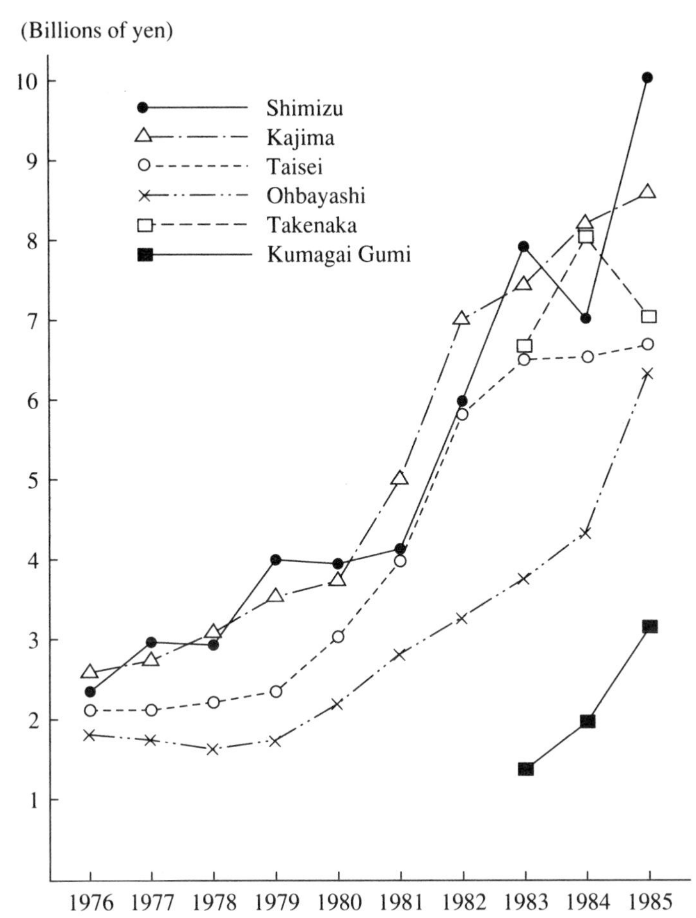

Source: Corporate annual reports for 1976 to 1985 [except for Takenaka Komuten and Kumagai Gumi, figures taken from *Nikkan Kensetsu Sangyo Shinbun*, a construction trade newspaper (June 23, 1986)].

Figure 7.2 Growth of R&D expenditures by the Big 6 (billions of yen)

nologies and foundation work techniques. Also, to expand residential space in big cities, there is a growing market need for high-rise residential buildings with proper seismic characteristics. Furthermore, as a small and mountainous country, Japan is expected to require many more structures in offshore, undersea, mountain, and aerial locations; contractors hope to take the lead in this field by accumulating the necessary technologies.

Compared to U.S. contractors, their Japanese counterparts have always maintained larger laboratories and research staff, partly because public regulations have made it difficult for private companies to commission research projects to universities and public research institutes and partly because there were few high-standard private research organizations in Japan. Their large research outfits have enabled Japanese contractors to be consistently active in R&D projects, but they are now beginning to reach a dead end in their attempt to use technology as a means to differentiate themselves from competitors.

Limitations of Differentiation

In the face of the prevailing price competition, contractors wish to use superior technologies to beat rivals, but it is increasingly difficult to achieve the needed differentiation by ordinary R&D efforts. All major contractors are keenly aware of the importance of research and development, and they allocate large amounts of equipment, staff, and money for it. However, even if one contractor takes an initial lead in a specific technology, others are quick to catch up so that all major general contractors are eventually on more or less the same level of technology. Also, except for a few sophisticated technologies, the technical gap between large and medium contractors has actually narrowed in both the construction and civil engineering fields.

Another reason for the difficulty in achieving technological differentiation is that because large contractors aim at all market areas to attract a maximum number of orders, their research targets and investments are too thinly spread. Yet another reason is that most of the technologies used in the construction industry today are time-proven technologies that serve existing requirements very well, and it is hard for new technologies to replace those traditional skills. The fourth reason is related to the civil engineering market, where 70% of orders are for public construction work. Because a number of construction companies are usually appointed to undertake each public work project jointly, the unique technology of a company required to accomplish the project is transmitted to the other members, lessening the differentiation effect.

Despite these difficulties in achieving technology differentiation, contractors are obliged to step up R&D activities for fear of being left behind in the R&D race, and this has ironically resulted in a me-too tendency by each contractor to acquire technologies that other contractors already have.

NEW PERSPECTIVES ON TECHNOLOGICAL DEVELOPMENT

The Necessity for Policy Redefinition

Under price competition, contractors strive to develop original technologies with the hope of increasing their appeal to prospective clients, but their expectations of technological development now appear to have reached perilous proportions.

This has happened because, in an effort to outbid competitors, contractors have often disregarded the essential purpose of technological development, which is to strengthen the company management. Technological development for differentiation is reaching a dead end, and it is time for contractors to reformulate their strategies from a management standpoint.

The technological development pushed forward to date was mostly carried out to maintain the same level as competitors, but contractors must redefine their strategies to emphasize their characteristic activities in research and development; that is, to differentiate the types of technologies that will be developed with special effort from the types of technologies that will receive only limited effort merely to maintain the technological status quo.

Unification of the different processes of company operations offers new research and development themes concerning methods of improving the productivity of the company as a whole. If technologies intended to differentiate the company from its competitors are considered as *offensive* technologies, those aimed at improving in-house productivity can be regarded as *defensive* technologies, and any effective technological development strategy should be based on a balanced combination of offensive and defensive technologies.

Developing Technologies to Boost Bargaining Power

The majority of technological development activities in the past were carried out within the borderlines of existing construction technologies, such as research into methods of reducing building costs. In the case of intelligent buildings, for example, most prospective technologies aim at the information and communications systems installed inside the building, but in reality, contractors are concentrating their R&D efforts on underfloor wiring systems, air conditioning panels, and other technologies closely associated with the structure of the building. This makes it hard for any contractor to master an innovative technology that will give the company a clear-cut advantage over its competitors.

A technical advantage is derived only from innovative technological development, and innovative technologies require the support of basic research and development activities in a broad area of scientific fields extending over a vast number of subjects. For this reason, licensing purchases from and technical collaborations with companies from other industries are important in order to achieve a speedy entry into a new market. Unless contractors implement radical technological development activities requiring a major decision-making effort by management, it will be difficult to obtain new technologies capable of serving as an irresistible lure to clients.

To develop truly innovative technologies, the company management must make crucial decisions to pinpoint the exact kinds of markets that will be developed as a result of such technologies. The management must also make major decisions on plans to inject large amounts of R&D funds and on a commitment to promote the relaxation of public regulations that might hinder the commercial application of new technologies. Though difficult, it is possible to acquire technologies that facilitate bargaining power. For example, a host of important new technologies have been developed, including the flexible steel structure technology, since the first super–high-rise building was erected in Japan in 1967.

The Bechtel Group of the United States is one engineering constructor that commands many attractive technologies serving as clinchers in negotiations with would-be clients. On the strength of its technologies, Bechtel operates throughout the world and, in 1985, rang up orders totaling $7.36 billion, or about twice the amount of orders posted by large Japanese contractors. Bechtel is noted for its diverse technologies spanning nuclear power plants, thermal and hydraulic power stations, oil refineries, pipe lines, mining systems, rapid transit systems, and hotels and commercial buildings. The basic approaches taken by this U.S. engineering constructor to achieve a superior technological capability are as follows.

1. Establishment of separate companies.
2. Acquisition of and joint ventures with other companies.
3. Capital investment in companies operating in strategic business fields.
4. Strengthening of in-house organization.

In addition, Bechtel recruits first-class engineers from General Electric, Westinghouse, and other major corporations. Although these steps may not be an exact model for Japanese contractors to follow, they do provide some meaningful hints. One hint is about the importance of a close linkage between technology and business strategies, which ensures that the company will allocate the necessary amount of resources to acquire specific technologies and will make a rapid entry into a new market once the necessary technology has been acquired. Another hint relates to a need for multifarious approaches to obtaining new technologies from outside the company. Although the most popular approaches in Japan have been in-house R&D activities and joint R&D activities with other companies, the Bechtel case suggests the availability of more diverse options if development strategies are considered as part of business strategies.

Increasing the Added Value: Offensive Technologies

Today it is virtually impossible to gain an unrivaled bargaining power from ordinary technological development efforts. Although the majority of R&D activities have been aimed at ways to reduce construction costs, it is time for Japanese contractors to reverse their way of thinking and aim their R&D efforts at increasing the cost performance, or the added value, of their products for the benefit of clients. In other words, contractors must replace their restricted technology differentiation approach with a more totalized differentiation approach combining marketing, engineering constructor, and development project strategies. Innovative technologies and services may not seem attractive to clients unless the contractor presents them as an attractive package. The top priority, therefore, should be placed on the development of technologies and services meeting the needs of clients.

Strengthening the Management: Defensive Technologies

In all industries, an important goal is to raise productivity as part of the effort to strengthen the company management, and the construction industry has to improve productivity through technological development. Japanese contractors have tried to raise productivity in the various stages of the construction process: the marketing, design, execution, and maintenance stages. The marketing department, for example, has developed real estate information and land-use evaluation computer programs; the design department has centered its research activities around CAD systems; the construction department is involved in such subjects as site management systems, process management systems, and robotics development; and the maintenance department pursues research themes such as external wall inspection systems and building diagnostic systems.

From a management point of view, however, the most important objective is to achieve higher productivity of the company as a whole, rather than to improve efficiency in individual processes, and the increasing availability of high-level information and communications technologies presents a great prospect for improving the productivity of the entire company by a remarkable margin. It has also become possible to quickly and reliably determine market needs by analyzing marketing information, and the results can be easily fed back to the design department to plan suitable products. Information from construction sites also can be passed on to the design department to help develop new construction methods.

FUTURE DIRECTIONS OF TECHNOLOGICAL DEVELOPMENT

With the importance of combining the offensive and defensive types of technologies in mind, Japanese contractors may try to obtain the following groups of technologies.

1. New frontier market technologies.
2. Software technologies.
3. Advanced technologies.

The New Frontier Market

Japanese industry is high on the idea of new frontier development, and the present plan to build a trans–Tokyo Bay highway through joint government-private efforts seems to be merely the tip of a massive iceberg. Since Japanese city areas are overcrowded and plagued by prohibitively high land prices, contractors feel that the demand will increase for the construction of offshore cities, undersea tunnels, and other marine structures in the near future. In 1986, the Ministry of Construction responded to these expectations by setting up a "new frontier committee" to define new frontier areas and to reflect ideas for new frontier projects in the ministry's administrative policies. To the delight of contractors, the new frontier market offers the following possibilities.

1. Creation of a giant construction market.
2. Application of new frontier technologies to existing design and construction technologies for land structures.

Including marine, underground, super high-rise, and space development, the concept of a new frontier market provides the principal theme of general contractors' technological development strategies geared toward the twenty-first century. Moves to acquire technologies for the new frontier market will result in innovations that can be applied to expanding the existing construction market as well. On a macroeconomic plane also, new frontier projects will help stimulate the domestic economy.

SPACE DEVELOPMENT. In April 1987, Shimizu Corporation set up a space development section within its main office organization. The contractor had earlier signed a technical collaboration contract with Starnet Structures of the United States. Since 1986, Shimizu and a number of other general contractors have been active members of a Japan Space Utilization Promotion Center formed by interested private businesses from all industries.

For Zenecons hoping to enter the space structure market, the first targets are those ground-based facilities for supporting space activities. Perhaps reflecting the increasing emphasis placed on the improvement of ground-based support facilities, the Japanese government has announced its intention to build a new space control center in Hokkaido, northernmost of the four Japanese main islands. In its "Underground Development Vision," the Science and Technology Agency of the government has revealed a plan to construct an underground facility for research on weightlessness.

Contractors see these developments as an opportunity to enter the space market, with a view to gradually expanding operations from land to space structures in the future. Once a sufficient number of space infrastructures are constructed, it will become possible to perform various research activities in space and to apply the acquired space knowledge to advanced construction technologies relating to materials, structures, and robotics engineering. In addition, space development must be carried out in close cooperation with the governments and industries of both Japan and foreign countries, and in this sense, space development strategies have a common interface with international strategies for Japanese contractors.

Development of Software Technologies

Software technologies may be developed to support full turn-key services and marketing activities. Full turn-key services provide not only conventional construction products and services but also other important associated services, enabling the contractor to present better project proposals to the client. If the contractor is to propose a housing structure plan for a fully automated plant after the plan for the production lines and production equipment has been decided, the contractor will face stiff price competition. On the other hand, the full turn-key approach is an attempt to win orders for not only housing structures but also production installations and all engineering services, such as the evaluation, design, procurement, installation, and trial operation of the production lines and equipment. Thus, the client benefits from the advantage of dealing with a single contractor, and the contractor is spared unnecessary price competition.

Before winning full turn-key orders, the contractor must develop certain technologies by a specific time and must be able to pinpoint the exact technologies that will become indispensable to attract full turn-key orders. Also, the contractor must determine the types of facilities for which full turn-key orders are expected to increase in the future, and must predict the potential market size for these facilities. Some of these technologies considered necessary to pull full turn-key orders in the future will be totally new to contractors, requiring them to launch R&D projects from scratch.

Marketing technologies are necessary for project planning, which determines the basic conditions for the design of structures and facilities. In the case of a hospital project, marketing technologies will be utilized in surveys of medical needs in the local market; in the formulation of a hospital income-and-expenditure plan; and in the determination of medical specialties, patient hospitalization capacity, and a pesonality character for the projected hospital based on local medical needs. These analyses will be conducted from a social engineering point of view, with an emphasis on analysis by computer simulation.

Because project planning requires accurate data and sophisticated model analysis, the contractor will need project planning specialists in each market field. Although many technologies used in site construction work are applicable to different types of structures—for example, collective housing and supermarkets (see Figure 7.3)—the number of interchangeable technologies sharply decreases in the initial stages of the construction project: the proposal, planning, and marketing

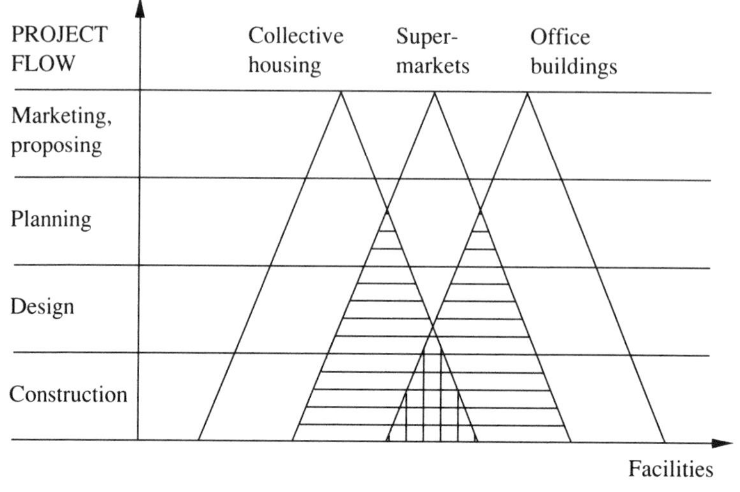

Figure 7.3 Project flow and interchangeability of technologies

stages. Put differently, these stages offer the contractor maximum opportunities to differentiate technologies from those of competitors. Although marketing technologies may not be subject to research and development in a strict scientific sense, contractors have yet to develop them into a general application system, so there is still room for contractors to improve marketing oriented technologies.

Accelerated Research and Development in High Technology

Advanced technologies will, of course, have great importance in Japanese contractors' future technological development. Through the research and development of high technologies, contractors will gain many opportunities to achieve innovations in their construction processes for the improvement of productivity (process innovation), and the ability to create new materials and structures (product innovation). High-tech research and development also facilitate the possibility that the contractors will be able to launch new businesses using newly developed technologies.

PROCESS INNOVATION. Many electronics and computer technologies are already being used in construction processes. These include structural computation, CAD, site management, process management, and construction robot control systems. Today there is a strong interest in artificial intelligence technologies, and since many construction activities are based on experience and personal know-how, artificial intelligence is considered the most suitable tool to automate these activities. Combining electronics with mechatronics and sensor technologies, artificial intelligence is expected to handle a wide array of basic white collar work, the selection of construction methods, diagnosis of cracks in concrete structures, project planning and design maintenance, thus enabling white collar workers to concentrate on higher-level jobs. In the development of CAD systems, a more efficient constructon of higher-quality facilities will become possible by linking a new CAD system with the CAD units used on construction sites and with a building repair-maintenance information system.

PRODUCT INNOVATION. As a successful example of efforts to develop new materials, Shimizu Corporation (together with Dai Nippon Glass Industry Company, a major sheet glass producer) has created "New Fiber Mesh," a new composite material of glass fibers and carbon fibers hardened with a synthetic resin and capable of achieving a strength greater than that of steel reinforcing rods and steel bars. The anticorrosive and nonmagnetic New Fabric Mesh can offer various strength levels to suit specific structural designs by changing the composite ratios of the two different fibers.

The use of this new material is expected to expand from current use in civil engineering work projects to such applications as use in coastal structures to prevent salt water corrosion. It is also lightweight, and rods and bars made of New Fabric Mesh can be carried by hand on the construction site. Moreover, this lightweight characteristic will significantly reduce the costs of structures. Although the use of New Fiber Mesh is limited at present because of high price and legal regulations, it is certain that this material will be mass-produced and put to extensive use (after safety tests) in the near future. The example of New Fabric Mesh suggests that the development of new materials provides many spin-off technologies and is an effective approach to acquiring innovative technologies.

TECHNOLOGICAL POTENTIAL AND FUTURE MANAGEMENT STRATEGIES

Today, the level of Japanese construction technology is at its highest point ever; the research and development activities of Japanese general contractors cover not only construction technologies, but also electronics, biotechnology, urban redevelopment, space development, and new frontier market development. Their employees boast a wide range of specialized knowledge in both natural and social sciences. Thus, Japanese contractors possess a great deal of technological potential in a wide range of fields. One important task will be to alter a tendency to accumulate technologies only for in-house use, and to use these technologies in diverse applications in the coming age of international exchanges, aging populations, and extending information and communications networks. The rapidly changing market environment under the strong influence of high technologies will provide increasing opportunities for contractors to utilize their vast reserves of technologies in the market.

Technological Development and New Business

When construction companies plan new business, they usually start from areas closely related with the main construction business line. A slight deviation from this approach is a recent increase in the number of subsidiaries set up by contractors to sell their technologies and resultant products. Technet, a subsidiary of Shimizu Corporation (an early example of such subsidiaries), sells solar optics systems, area calculators, and other merchandise. These are the by-products of R&D activities, and their unit prices are low. In contrast, the aforementioned New Fiber Mesh was developed from the beginning for sale and manufacture through a subsidiary because of the large market potential of the projected new material. Preferably, contractors should develop each technology in such a way that a business can be formed around that technology.

Expansion of Research and Development Services

Visitors to the laboratories of large Japanese contractors are amazed by the high-level research facilities. The visitors are surprised that many of the research facilities in these laboratories are not directly connected to construction. Such facilities include industrial clean rooms, bio-clean rooms, and acoustic, hydraulic, and wind tunnel testing laboratories. As in psychological experiments, some contractors will probably build facilities to determine ideal human environments, actually using human subjects in the foreseeable future. At the research center of Shimizu Corporation, for instance, the majority of researchers are specialized in areas outside the construction and civil engineering fields proper. Some observe the life of microorganisms in bioreactors; others measure the performance of dust-free clothing worn inside semiconductor plants; and others work hard to develop simulation programs for unmanned conveyor systems used in fully automated factories.

Since contractors handle a variety of structures, their research facilities must be large, so the number of technologies turned out by their researchers is correspondingly large. Utilizing their ample R&D facilities and staff, contractors may eagerly invite R&D projects from other companies on a commission basis in the future, a realistic idea considering the vast research resources available.

Technology Exports and International Cooperation

For overseas observers, Japanese construction may be most closely associated with antiquake technologies and historical temple and shrine architecture. Although the

Japanese construction industry imported a large number of technologies from the West in the postwar period, it has now developed its own technologies that may be exported to other countries. On the strength of their high-level construction technologies, it is possible for Japanese contractors to sell their expertise in the building of space and other high-tech structures because sophisticated technological activities require sophisticated structures. The contractors can also take orders for big international projects. Japanese construction technologies, especially those for building city infrastructures, may be transferred to developing countries as part of international cooperation projects. Contractors should regard contributions to international cooperation as an important part of their operations.

The Growing Importance of R&D Management

Japanese contractors will continue to uphold research and development as a cornerstone of their business strategy. Since R&D is a key to corporate growth, effective management policies must be formulated concerning the determination of R&D target areas, employment of engineers, and allotment of financial resources. If R&D targets are selected only from a technical standpoint, resultant technologies will probably not meet client needs and will be rejected. R&D themes should be selected on the basis of extensive information gathered about social and client needs.

A challenge in maintaining research staff is to deal with the problem of researchers unable to keep pace with the rapid advance of technology. The existing practice of employing young researchers immediately after graduation from the university until retirement age, gradually training them to become capable researchers, is becoming too costly to the company. Japanese contractors will have to recruit a greater number of mid-career researchers than they are now doing, although this is accompanied by the risk of losing talented researchers to other recruit-minded companies. The importance of research staff management will further grow as a crucial link between corporate management and technological strategies.

Shimizu Corporation for example, spends about 10 billion yen ($67 million) per year for research and development, and will continue the same investment level in the future. Despite the massive amount of money involved, there is no simple equation by which the management can evaluate the fruit of R&D investments. Nevertheless, the management can put research and development in the center of its overall strategy by trying to assess the specific contributions from each R&D project. In sum, the effective management of R&D themes, personnel, and financial resources is an essential ingredient for the growth of the company.

Nagai River Bridge, Kan-etsu Expressway (Japan) Zenitaka Corporation

CHAPTER 8

FINANCIAL STRATEGY

Several large Japanese contractors have established financial subsidiaries abroad in recent years: Kumagai Gumi in Australia and Taisei in the Netherlands in 1985, and Kajima in Hong Kong and Shimizu in London and New York in 1986, to name a few. Domestically, Kajima established a subsidiary, Kajima Lease, and Shimizu set up SC Finance Engineering, in 1986. These moves are closely linked with changing construction environments and are part of the corporate strategies of contractors. Finance is a vital element of corporate strategy, together with personnel and technology. The effectiveness or ineffectiveness of financial strategies, to a great extent, determines the profit performance, the management strength, and the rise or fall of the company.

CHANGES IN THE FINANCIAL STATUS OF CONSTRUCTION COMPANIES

The financial conditions of Japanese contractors have suffered a dramatic change over the past years, and the financial data of Shimizu Corporation for fiscal 1972 (immediately before the first oil crisis), fiscal 1982 (10 years later) and fiscal 1985 (latest data available) is typical of this situation. A notable development in Shimizu's financial status is a sharp rise in the costs of construction in progress, from 32.0% of total assets in fiscal 1972 to 48.9% in fiscal 1982 (see Table 8.1). During the same 10-year period, the percentage of overseas orders in Shimizu's total orders awarded ballooned from nearly zero to 10.6%, and became a main cause of the sharp rise in the costs of construction in progress.

Another noticeable development is the brisk growth in the sales of real estate items, from 3.2 to 8.7% of total sales during the same period, reflecting the company's increased activities in the real estate development business. Also notable is the growth in the company's investments in the equity shares of subsidiaries from 0.7% of total investments in fiscal 1972 to 1.3% in fiscal 1985, indicating an increased effort to develop new business through the operation of subsidiaries. Shimizu's gross assets expanded 180% between fiscal 1972 and 1985, and this gain was accompanied by a marked transformation in assets makeup as the company forged ahead with strategic operations such as overseas projects, development projects, and new business development operations.

Diversified Methods of Fund Procurement

In Shimizu Corporation, as in other leading Japanese contractors, an increasing number of different fund procurement methods have been used to keep pace with enlarging assets. Between fiscal 1972 and 1982, the balance of construction

Table 8.1 Financial Data of Shimuzu Corporation (Unit: Billion Yen)

	End of March 1973		End of March 1983		End of March 1986	
	¥	%	¥	%	¥	%
Current Assets	287	79.9	773	85.6	838	84.5
Cash on hand in banks	87	24.0	114	12.6	111	11.2
Cost of construction in progress	115	32.0	442	48.9	399	40.3
Real estate for sale	–	0	29	3.2	87	8.7
Other current assets	86	23.8	189	20.9	241	24.3
Fixed Assets	72	20.1	130	14.4	154	15.5
Investments	47	13.0	90	9.9	114	11.5
(of which, stocks of affiliates)	(4)	(1.2)	(7)	(0.7)	(13)	(1.3)
Other fixed assets	26	7.1	41	4.5	40	4.1
TOTAL ASSETS	360	100.0	904	100.0	992	100.0
Current Liabilites	249	69.1	709	78.4	701	70.6
Short-term borrowings	57	15.7	57	6.4	83	8.3
Advances received on construction in progress	119	33.0	461	51.0	368	37.0
Other current liabilities	73	20.4	190	21.1	251	25.3
Long-term Debt	29	8.1	55	6.1	134	13.5
Bonds	0	0	0	0	30	3.0
Long-term borrowings	26	7.1	19	2.1	67	6.7
Other long-term debt	3	0.9	36	4.0	38	3.8
Shareholders' Equity	78	21.6	140	15.5	157	15.9
Common stocks	16	4.4	36	3.9	36	3.6
Other shareholders' equity	62	17.1	104	11.5	122	12.3
TOTAL LIABILITIES AND EQUITY	360	100.0	904	100.0	992	100.0
Balance of construction in progress	3		19		–32	
Orders received	446	100.0	1,042	100.0	989	100.0
(of which, overseas orders)	(0.1)	(0)	(110)	(10.6)	(64)	(6.5)

Source: Annual reports of Shimizu Corporation for fiscal 1973, 1983 and 1986.

in progress* continued to improve, posting a total improvement of 16 billion yen ($110 million) compared to 10 years earlier, thanks to a favorable condition of payments from clients (see Table 8.2). Accordingly, during this 10-year period, Shimizu was able to boost its shareholders' equity by 62 million yen ($410 million) through capital increases. As a result, Shimizu was able to increase its business assets including real estate pieces for sale. In addition, the company was able to curb its debt while boosting cash deposits.

When fiscal 1985 is compared with fiscal 1982, however, the balance of construction in progress sharply worsened to a deficit of 51 billion yen ($340 million), partly because of a decrease in advance payments for public works projects and partly due to harsher payment conditions created by money-conscious clients. At the same time, to continue brisk purchases of real estate items for sale and to beef up fixed assets, Shimizu increased long-term and short-term debt,

*The balance of construction in progress is the difference between revenues from clients and payments for subcontracting and other construction services. The costs of construction in progress account for a large portion (about 40% for Shimizu) of total assets, and advances on construction in progress are also proportionally large in total liabilities and equity, because of the prevalent "completed contract accounting method" under which net sales are recognized only when the constructions are completed. For this reason, the balance of construction in progress deeply affects the cash flow of the contractor. If the balance shows a constant surplus, the contractors can enjoy an easy cash flow.

Table 8.2 Working Assets and Fund Raising of Shimizu Corporation (Unit: Billions of Yen)

	Fiscal Years 1972–1982		Fiscal Years 1983–1985	
	Working assets	Raised funds	Working assets	Raised funds
Cash on hand and in banks	27.2			2.8
Real estate for sale	29.2		57.3	
Fixed assets	57.9		23.7	
Balance of construction in progress		16.0	51.2	
Bonds		0		29.9
Long- and short-term borrowings	5.8			72.5
Shareholders' equity		62.4		17.4
Others		41.7		9.6
TOTAL	120.1	120.1	132.1	132.2

Source: Annual reports of Shimizu Corporation for fiscal 1972 through 1985

spent some of its reserve funds, and issued a corporate bond for the first time in an attempt to procure a cheaper fund.

Most Japanese contractors are now experiencing a worsened balance of construction in progress, while expenditures for development projects and the loans to and investments in affiliated companies have increased over the past three or four years. As a result, the contractors have resorted to issuing ordinary, convertible, or warrant bonds in succession, in addition to increasing debt. As contractors try harder to obtain cheaper funds for new business development activities, they will utilize diverse methods of fund procurement, and the importance of strategies concerning project finance, group financing, and financial risk management will grow further.

FINANCE STRATEGIES

While, in the face of stiff competition, contractors make persistent efforts to create orders by improving their project proposal and planning abilities, they are also increasing the use of finance as a tool to attract orders. In this regard, their financial activities can be divided into four types.

1. Project financing, whereby the contractor introduces prospective clients to interested financial institutions.
2. Guarantee of loans borrowed by clients.
3. Investment in construction projects.
4. Establishment of finance and leasing subsidiaries.

Project Financing

A realization of large overseas construction projects depends heavily on the availability of low-cost funds, and a fund shortage can result in a suspension or delay of the project. Accordingly, a contractor can attract orders by providing a project financing service, such as receiving credits from government organizations and persuading banks, insurance companies, and other investors to join an international consortium as financing members. Kumagai Gumi is especially noted for its active involvement in project financing. For example, this internationally minded contractor was able to arrange an international syndicate loan of 2.7 billion Hong Kong dollars (54 billion yen) for the construction of a second undersea tunnel in Hong Kong, hiring Shearson and Lehman Brothers as an adviser. The loan was particularly attractive, with low interest on the prime rate plus 0.75% in the first three years and long repayment terms of 15 years for road construction and 18 years for railway construction. In all likelihood, other Japanese contractors will strive to beef up their project financing capabilities. To this end, they must establish channels of communication with financial institutions all over the world and upgrade their abilities to collect and analyze information on world financial developments, international construction projects, and country risks.

Guarantee of Loans

Large general contractors have increased their guarantees of loans by other companies (see Table 8.3). Contractors have long offered loan guarantees to cash-deficit clients, but recently they are consciously using loan guarantees as an effective means of promoting orders. While Kajima and Takenaka provide loan guarantees mostly in the domestic market, Kumagai Gumi is rapidly increasing its loan guarantees for overseas projects, recording, as of the end of September 1986, a total loan guarantee of 250 billion yen ($1.7 billion), of which 80% is denominated in foreign currencies. The prevailing prediction is that loan guarantees by Japanese contractors will further expand in both domestic and foreign markets.

Investment in Equity Shares

There is increasing demand from clients for investment in equity shares by contractors, together with requests for project financing and loan guarantees. Such investment is now regarded as an effective instrument for promoting orders. As the government outlay for public construction work is low under tight budgetary conditions, both the national and local governments in Japan have begun to invite private capital, planning know-how, and management concerns to undertake big public projects as well as urban and regional development projects. In the New Kansai International Airport and Trans-Tokyo Bay Highway projects, for example, large general contractors have invested by buying equity shares in the project

Table 8.3 Outstanding Amounts of Loan Guarantees by General Contractors (Unit: Billion yen)

	3 years earlier		Latest data year		
	Amount	Date	Amount	Share*	Date
Kajima	19.5	Nov. 30, 1983	38.2	90%	Nov. 30, 1986
Takenaka	88.6	Dec. 31, 1982	95.9	100%	Dec. 31, 1985
Kumagai Gumi	26.7	Sep. 30, 1983	249.7	20%	Sep. 30, 1986

* Share of loan guarantees provided in Japan.

Source: Annual reports of Kajima, Takenaka, and Kumagai Gumi, 1987

administration companies. Similar investments have been observed in an increasing number of regional development projects.

Investments in equity shares are also on the rise on the international front, led by Kumagai Gumi. This enterprising contractor invested in a 100 billion yen ($670 million) Australian project to redevelop Adelaide Train Station and the surrounding district in 1983, and today is engaged in a number of projects in the United States and Europe as an owner of joint ventures or a party in partnerships with local developers. In the future, Kumagai Gumi hopes to operate or sell the projected buildings for a handsome profit. Stimulated by Kumagai Gumi's moves to capture orders through investments in overseas markets, Japanese contractors are looking to the possibility of cultivating orders through similar investments in the domestic market.

Finance and Leasing Companies

In recent years leading contractors have set up finance and leasing subsidiaries designed primarily to provide construction funds to clients on behalf of the contractors. By mobilizing these specialized subsidiaries, the contractors can handle loans, loan guarantees, and factoring that are too risky or involve excess complications for the contractors. The subsidiaries not only earn profits from the financing business, but also have the synergistic effect of helping their parents to win orders, and even medium-sized general contractors are beginning to set up similar subsidiaries. Contractors as a whole are now aware that they can no longer get along merely by taking the construction risk but must also shoulder other risks peripheral to construction. Consequently, they must develop the skills needed to utilize various methods of sharing financial risks with clients, such as lending, project financing, loan guarantee, and investment.

GROUP MANAGEMENT STRATEGY

Since the low-growth economy has proved to be a permanent phenomenon, all industries are striving to increase their business lines to achieve higher growth. The construction industry is no exception. Although it is feasible to launch new business operations through in-house special assignment teams, or through the business operations department, the common strategy of contractors today is to establish a special-task subsidiary or a joint venture for each new business. This is especially important for new business not related to construction because subsidiaries are less liable to follow the construction-centered thinking of the main office. In addition to the establishment of subsidiaries, contractors are certain to increase the number of their affiliates through mergers and acquisitions, a currently popular idea in the

Japanese business community. They will also devise a group management strategy designed to achieve the prosperity of the entire group while allowing maximum freedom to each group member.

In-Group Financing

Contractors now realize the growing importance of a more effective management and procurement of funds and bolstering in-group solidarity as part of the group strategies. Financial subsidiaries have recently been established by major contractors in Japan, Britain, Hong Kong, and elsewhere. One reason for doing so is to acquire the most advantageous funds and find the most favorable instruments for fund management in the international money market in the interest of their group members. Since the parent company is expected to step up the use of subsidiaries and other group member firms in promoting development projects and new business operations, the importance of group financing will further increase.

Merger and Acquisition

Although general contractors have commonly pushed their business diversification strategies through subsidiaries and affiliates and joint ventures with other companies, a number of construction companies now utilize a merger and acquisition approach as an effective method of diversification. Misawa Homes, a large specialized prefabricated home builder, is a forerunner in this concept, having taken over a succession of companies since 1983 by acquiring their equity shares. Through a company acquired in 1983, Misawa now undertakes a VAN (Value-Added Network) business serving small housing and realty firms. Misawa is also trying to develop ceramic materials for home construction through a company acquired in 1984, and a home greenery business through a company bought in 1985.

Although Misawa acquired companies at a pace of one a year from 1983 to 1986, it purchased three companies in the first three months of 1987, including a firm engaged in research into home energy systems and another in the development of leisure and resort areas. Misawa is careful to buy companies from electronics, new materials, biotechnology, and other high-tech fields while placing top priority on housing-related business lines. Fully cashing in on today's low interest rates, it collects corporate acquisition funds by borrowing from banks and issuing bonds with warrants.

Except for Misawa and a few others, mergers and acquisitions (M&A) have been rare in the Japanese business community. One reason for this is that companies had little need for M&A because their business expanded rapidly during the high economic period. Another reason is that many Japanese companies own equity shares of other companies with the intention of strengthening business relationships on a long term basis. The third reason is the negative image of mergers and acquisitions as a piratical act on an unfortunate company. In a slow economy, however, the Japanese are beginning to see mergers and acquisitions as a time- and money-saving way of business diversification, and the negative image of M&A is fading fast as many books on mergers and acquisitions in the United States and Europe have been published.

As a result, the number of companies with active M&A experience has perceptibly increased. Minebea Company, the Japanese leader in bearing production, is particularly famous for take-over bids in the United States. Misawa Homes, on the other hand, has concentrated on domestic mergers and acquisitions. Because the mergers and acquisitions practiced so far have been between friendly companies, the Japanese public has responded with understanding, and it is most likely that mergers and acquisitions will continue to increase at a steady pace.

REDUCING RISK

As Japanese general contractors expanded the realm of their business activities, they encountered new risks: foreign exchange risks in overseas projects, and risks involved in development projects and new business ventures. These risks must be hedged to the maximum degree.

Hedging Foreign Exchange Risk

In the wake of the first oil crisis in 1973, general contractors made inroads in Middle Eastern and Southeast Asian markets, but in some instances they suffered a loss due to a lack of experience in overseas construction activities. Foreign exchange losses frequently occurred because the overwhelming majority of construction contracts were denominated in the U.S. dollar or local currencies at that time. Nevertheless, contractors now implement a variety of hedging measures against foreign exchange risks. Although these measures do not yield exchange gains at a time of yen depreciation, they do eliminate exchange risks.

EXCHANGE CONTRACTS. Forward exchange, in which foreign currencies are transacted at fixed future rates, are widely used as a risk-hedging measure because

1. It is easy to take part in the transaction.
2. No funds are required.
3. It is an off–balance sheet transaction.
4. Alteration and cancelling of transactions can be flexibly carried out.

IMPACT LOANS. Since contractors are usually paid in the local currency or in the U.S. dollar for their services, they incur an exchange loss when the Japanese yen appreciates. This possible loss can be hedged by obtaining impact loans (foreign currency loans with no spending specifications), which will yield a gain when the yen appreciates.

BONDS WITH WARRANTS. Bonds with warrants give the owner the right to purchase the issuer's stocks at predetermined conditions while keeping the bond in force. On the other hand, convertible bonds expire when they are converted into stocks. Thus, bonds with warrants can serve as a hedge against yen appreciation. The interest rates of these bonds, however, are higher than those of convertible bonds.

EXCHANGE CONTRACT WITH AN OPTION. This type of exchange contract allows the user of yen futures to cancel the futures and buy yen at the ongoing rate in the event of yen fluctuations. In this way the user is permitted to gain from yen depreciation while hedging against yen appreciation. Although this option demands an extra cost, users consider the option to be a kind of insurance.

FACTORING. Factoring is a credit purchasing operation consisting of three basic processes: (1) the factoring company buys the book credits and bills from the customer; (2) the factoring company collects the payables; and (3) when the payables are dishonored, the factoring company shoulders the entire risk. For companies with book credits, factoring gives cash without having to collect from creditees. Maeda Construction took its book credits (denominated in Hong Kong dollars) from a Hong Kong subway construction project to a factoring bank when the Hong Kong dollar depreciated against the Japanese yen. In exchange for the sale of the book credits, Maeda was also able to receive an amount of cash in yen from the local bank. This eliminated the exchange risk of the Hong Kong dollar and at the same time enabled Maeda to earn interest through the management of the yen fund. Apparently, Maeda chose factoring over an impact loan as the best method of hedging exchange risks, because the interest rates of impact loans are too high for time-consuming construction projects.

OVERSEAS FINANCE SUBSIDIARIES. To avoid foreign exchange risks, foreign currency proceeds from overseas projects may be deposited in overseas finance subsidiaries, and when funds are required to run overseas projects, the foreign currency deposits can be supplied. The parent company can not only earn interest on these deposits but also expect a dividend from the financial subsidiaries after their operation becomes profitable.

Hedging Investment Risk

"High risk, high return" is the keynote in many types of investment activities. Contractors are increasing their investments as they strive to increase their lines of business, but they must make every effort to minimize predictable or preventable risks. Investment risks are divided into emergency risks and business risks; in particular, risks involved in overseas investments are hedged from a long-range perspective little by little over a long period of time. Accordingly, the conditions of host countries, management environments, and other important risk-related developments must be accurately predicted over a long range. To this end, the contractor may build an in-house information network using branch offices, subsidiaries, and other group companies in order to establish channels of information with financial institutes and research organizations in Japan and abroad. The contractor must, of course, have the ability to analyze the information collected.

Another method of reducing investment risks is to subscribe to overseas investment insurance and enter partnerships with local investors to disperse the risks. Also, the contractor must produce a contingency plan to be able to act correctly and quickly in the event of an emergency. Although Japanese construction companies have noticeably improved their exchange risk management through experience in foreign exchange losses, they have not incurred any major losses from their investments as yet. For this reason, they are less prepared for investment risks, and must make the utmost effort to prevent losses by studying the case histories of companies from other industries. The time in which contractors shoulder only construction risks is ending, as they ambitiously try to broaden their business realm through international operations, new business ventures, bolstered engineering constructor capabilities, development-type construction projects, technological innovation, and improved financial activities. For Japanese contractors, a new age has dawned—an age of high risks, but also of determination to grow by dealing with risk problems. The fundamental solution lies in the ability of contractors to alter their company structures to better suit the coming new era.

Kinki Nippon Railway Higashiosaka Line, Ikoma Tunnel (Japan) Ohbayashi Corporation

EPILOG

BECOMING A STRATEGIC COMPANY

THE MOST EFFECTIVE STRATEGY MIX

How can these six strategies be used by Japanese construction companies to achieve stable growth? How should the contractors select and mix these strategies to actively create a favorable business environment, rather than passively try to adapt to the changing environment? These six strategies—transnational expansion, new business development, integrated engineering construction, development projects, technology development, and financial management—are merely the warps of a fabric, and we must produce a complete fabric by weaving in the woofs of discussions on how to implement these strategies. These strategies rarely function independently but, in most cases, interact in an organic, dynamic manner (see Figure E1).

Sun-tzu, an ancient Chinese strategist, wrote: "There may be only five different tones, but all of their variations must be listened to. There may be only five different colors, but all of their variations must be seen. There may be only five different tastes, but all of their variations must be tasted. In military strategy, there are only orthodox and unorthodox strategies, but all of their variations must be studied. " Here, variations include mixed effects of two or more elements.

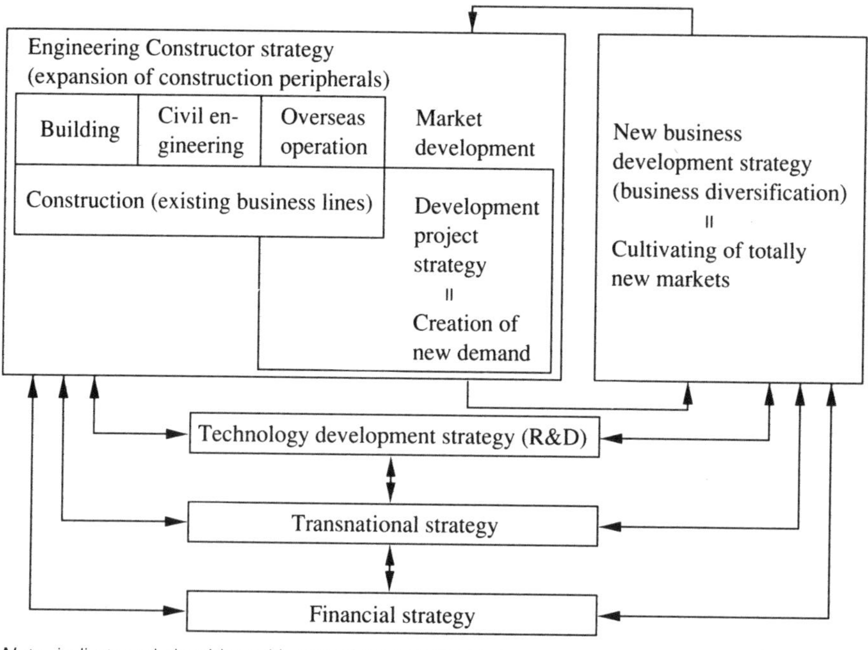

Note: indicates relationships with mutual synergistic effects.

Figure E1 Interrelationships of six strategies

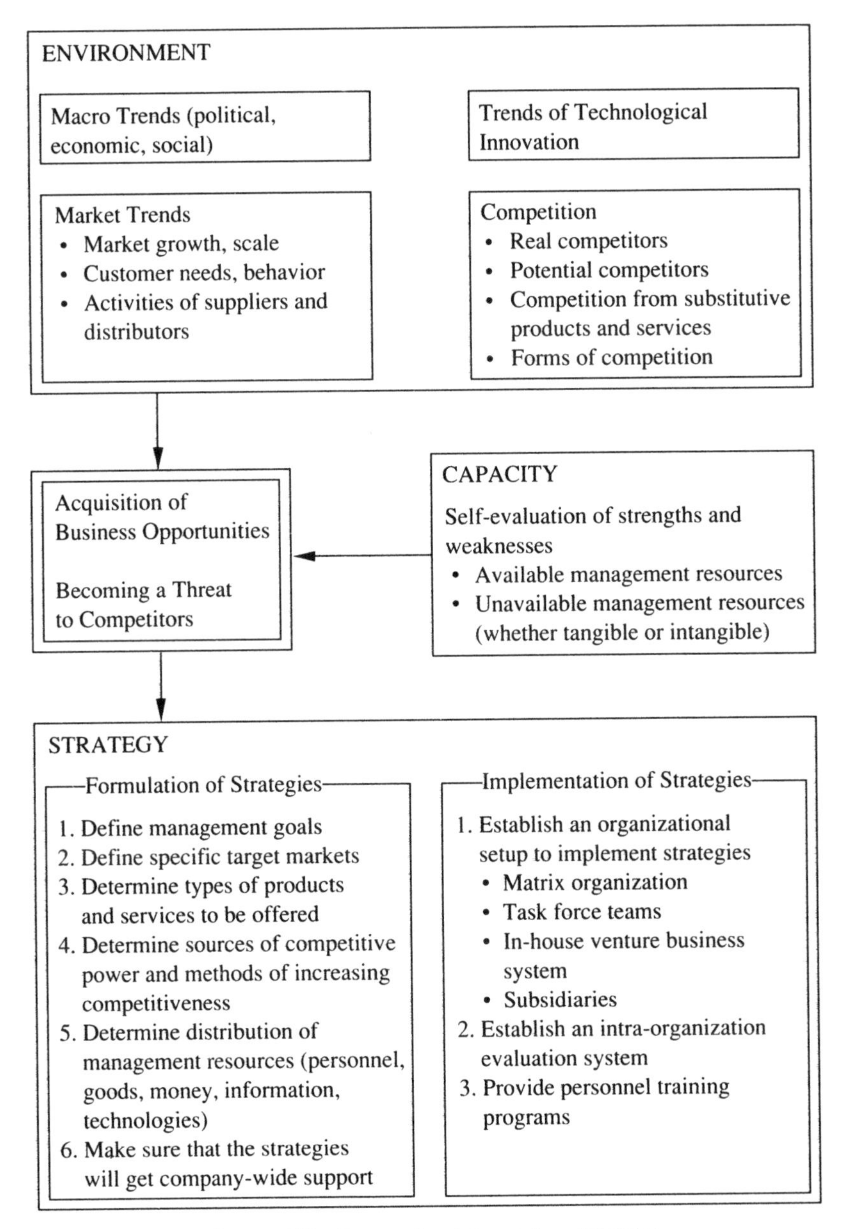

Figure E2 Factors for determining strategies

There is no ultimate strategy mix for contractors, but an unlimited number of strategy variations exist depending on business domain (management goals, product types), business environment (political, economic, social, technological, competitive), and business capacity (strengths and weaknesses), as outlined in Figure E2. To stimulate strategy mix, different approaches must be applied as woofs which, together with the six strategy warps, will produce a complete fabric.

PROPER APPLICATION OF STRATEGIES

Competitive strategies may take either of two major directions: a head-on collision with rivals for a greater share of the market, or a move into a field where there is little or no competition (Figure E3).

Strategies for Gaining Superiority

An important part of the effort to gain superiority over competitors is to identify the most powerful existing and future competitors. This prevents the use of incorrect strategies on irrelevant competitors. Once competitors are clearly identified, the company can fully evaluate the tools needed in the competition. Generally:

- The first need is to differentiate the products by quality and functions.
- The second is to differentiate services, including brand images, distribution networks, payment conditions, and after-sale services.
- The third is to differentiate prices, which in most cases means to lower prices.

In short, quality, speed, efficiency, and economy are the necessary ingredients to beat competitors, and all of the six strategies for contractors are the tools to obtain these ingredients under optimum conditions.

Strategies That Reduce Retaliation

Usually in the business world, competitors will put up a stiff retaliatory fight against a threatening force. Such a collision could become a lengthy, grueling affair, and it is advantageous to employ strategies against which competitors will find it difficult to retaliate. One way is to forestall the competitors' retaliatory feelings by creating a situation in which little or nothing will be gained by retaliatory efforts. Classical examples of this tactic are

- Introduction of a substitutive product on the market.
- Implementation of a price discount policy in a certain locality.
- Aggressive plant and equipment investment aimed at large market potential.
- Deployment of a large, increased number of sales workers.

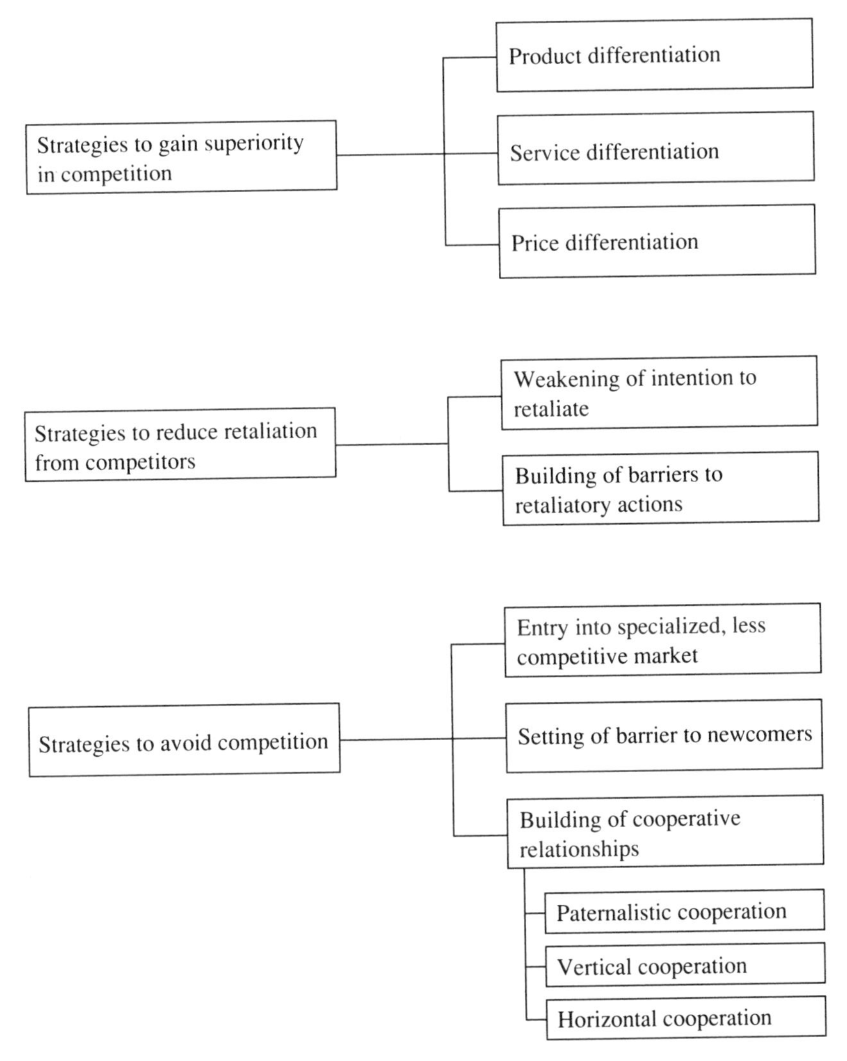

Figure E3 Strategies as woofs of strategy mix

In all cases, if the retaliator takes steps similar to those taken by the newcomer, it can be hurt by the disruptions in the market.

Another way to reduce retaliation is to set up barriers such that competitors cannot retaliate even if they want to. For example, the newcomer may attempt to stop the supply of important business resources to competitors by cornering machinery, materials, and skilled workers. Kajima Corporation has acquired a patent right on an emergency broadcast system using cable television networks in order to bar competitors from developing similar systems. Monopolistic seizure of certain distribution channels is another method of raising a barrier against competitors' retaliatory efforts.

Strategies That Avoid Competition

Despite all, avoiding competition is the most effective strategy, and consists of three approaches, as shown below. These approaches indicate that competitive strategies and cooperative strategies have a two-sided relationship.

ENTRY INTO LESS COMPETITIVE MARKETS. A plausible way of avoiding competition is to enter a less competitive or totally competition-free market area, as in the niche strategy employed by many small or specialized companies. This strategy, nevertheless, contains a dilemma in that if a company succeeds in its carefully selected market, new competitors will likely spring up from outside. To be successful in this strategy, the company must find markets where customer needs have not been fully met, consider effective ways to advance into such markets (after studying why other companies have not entered these markets), and provide protection against the anticipated entry of competitors (by setting up barriers or by selecting a market with high natural barriers to entrants).

BUILDING BARRIERS TO NEWCOMERS. These barriers are divided into two groups: the management resources type spanning patents, equipment capacity, technological levels, and distribution channels; and the regulation type including regulations on foreign companies, import activities, and large retail store operations. The barrier setter must remember that because of the quickened speed of technological progress and the shortened life of each technology, barriers can easily cease to be effective through the passage of time, and unexpected competitors can suddenly emerge. Today, the appearance of competitors should be regarded as inevitable, and the forerunner can only delay the time of competitor entry.

COOPERATION STRATEGIES. Although cooperative strategies may not seem to fit with competitive strategies, the importance of cooperation is increasing in today's interindustry business operations. While competition remains the ultimate driving force in the market, cooperation can also bring benefits to companies by eliminating excessive competition, reducing overlapping costs, teaming to develop common technologies, and joint efforts to promote market growth. There are three types of cooperation strategies.

1. The first is paternalistic cooperation, in which the large company provides the small venture business with the necessary funds, goods, and facilities in exchange for the joint development of the venture business's technologies into saleable

products. Investment in venture capital companies and capital participation in productive subcontracting companies are two examples of paternalistic cooperation.

2. Second is vertical cooperation, where the large company and the small venture business form an equal partnership in conducting research and development, procurement of machinery and materials, production, sales, or other steps on the vertical commercialization ladder. In the great majority of cases, large companies commercialize products using the technologies of small venture businesses through vertical cooperation.

3. Third is horizontal cooperation, where two large companies, or two small venture businesses, or a large company and a small venture business combine their technologies and experience to commercialize products or services. The construction and management of nursing homes for the elderly by the team of Shimizu Corporation and Beverly Enterprises of the United States belongs to this horizontal cooperation type.

THE PURSUIT OF *VARIATION* MANAGEMENT

The best mix of strategies is achieved when the external factors (customer needs, socio-econo-political trends, technological innovations, competitors' activities) and internal factors (management resources, i.e., personnel, goods, money, information, technologies) are combined in the quickest, most efficient, and most economical way. In this sense, strategy comes first and modifies organization, but it is also true that organization modifies strategy because organization, a collective body of human beings, always requires strategies to reflect human behavior.

In seeking the best mix of strategies, a point to consider is the observation that strategies that seem effective in the short term may not be effective over time. In the past, a leading steam locomotive manufacturer continued to pour most of its management resources into the development of more efficient steam locomotives and eventually disappeared from the market as they were gradually replaced by diesel trains and then by electric motor trains. In this case, an effective strategy from the short-term standpoint proved destructive in the long range.

Another point to consider is that there are always trade-offs in the combination of various strategies, where a bolstering of one strategy weakens another. Plots of land purchased for future development projects may become a burden from the financial strategy standpoint. When different strategies have conflicting interests, the company must determine the compromise point.

The third point to consider is the formulation of strategies capable of responding to changes in the surrounding market. Until recently, the most important challenge for the majority of construction companies was energy saving, but today the emphasis is shifting to intelligent technologies, compelling the companies to revise their technology development and other strategies.

The best strategy mix is achieved when the combined strategies produce the maximum amounts of ripple and synergistic effects. The idea of variations pointed out by Sun-tzu is highly relevant and should not be disregarded. In fact, all companies worthy of the title of "strategic company" accept the concept of variation and even try to create variations in their strategies.

Invariables in Variations

A strong and close relationship exists between society and the construction industry; structures built by contractors serve their users and dwellers for dozens of years or more, influencing people's activities and behavior to a significant extent and even affecting cultural conditions on the regional and national levels. The likelihood is increasing that such structures will be built not only on the ground but also underground, on offshore sites, and in space. Contractors are always anxious to construct buildings capable of satisfying their users over a long period of time, and long-term requirements cannot be satisfied if contractors are fixed on management strategies focused on making quick profits. Myopic approaches eventually yield adverse results, so contractors must show farsightedness in their strategy formulation.

MEETING THE INTERNATIONAL AGE

The Japanese industrial structure is undergoing a major change triggered by a sharp appreciation of the yen. The automobile, consumer electronics, and most other manufacturing industries have already established their production bases abroad, and the outbound movement is continuing. The emphasis in the nation's industrial policy is being switched over from heavy dependency on export activities to dependency on domestic market expansion. As part of this, interest in the Trans-Tokyo Bay Highway and other large-sized infrastructure construction projects is surging, and Japanese and overseas contractors allike are enthusiastic about this and other projects.

Growing Interdependence

Vast amounts of funds are being pumped out from Japan to the United States and other parts of the world, due to Japan's trade surplus and interest rate differentials. The flow of money is also stimulating the flow of technologies and people. Reflecting the stepped-up money management activities of Japanese companies, money from Japan is often spent on businesses and real estate in foreign countries. The establishment of an increasing number of overseas production plants by Japanese manufacturers will expand the transfer of Japanese technologies and contribute to the economic development of the host countries. The interdependence between Japan and other countries will therefore further increase, and there will be a greater need for co-prosperity between Japan and other countries. In the spirit of reciprocity, Japan must further expand its market access to foreign companies.

Interdependence with other countries is felt on the company level, but friendships with other countries cannot be achieved only by technological and economic considerations. The basis of any friendship is an understanding of different living conditions, history, cultures, and value judgments in different countries. These differences are particularly noticeable in the world of construction, where doors, for example, are pulled or pushed open in the West while they are slid open in Japan. The ideas of space are different, and accordingly, work procedures and management methods on the construction site differ widely between Japan and other countries. For construction activities to cross national borders, the contractor must achieve an understanding of the host country, and in this regard, the entry of foreign contractors in Japanese construction projects will help promote mutual understanding between Japan and other countries.

Improvement of Infrastructures for the Twenty-First Century

Japan now accounts for 10% of the world GNP, and its per-capita production value has reached the U.S. level. Yet Japanese housing, road, sewage, and other infrastructure conditions still lag far behind those of the United States and European countries. An estimated 1,000 trillion yen ($6.6 trillion) are required during the remainder of this century to improve, maintain and repair Japanese infrastructures. As Britain, France, and the United States have done, Japan should improve infrastructures while its economy is strong, and the improvement speed is predicted to accelerate as a result of the recent policy shift to the stimulation of the domestic economy. The undersea tunnel linking the northernmost Hokkaido Island to the main Honshu Island was scheduled for completion in the spring of 1988, and trans-sea bridges over the Seto Inland Sea are under construction. In addition, as of this writing, the New Kansai International Airport Project and the Trans-Tokyo Bay Highway Project have just started.

To smoothly carry out these construction projects, there will be a growing need for cooperation among Japanese contractors and also between Japanese and foreign contractors. Consequently, increased amounts of funds and technologies are predicted to flow into Japan, as foreign contractors step up their efforts to penetrate the Japanese construction market in search of greater business opportunities. Given this prospect, Japanese contractors are required to prepare business strategies for cooperation with foreign contractors, which will be effectively mixed with their competitive strategies in the coming years.

REFERENCES

1. *A Vision of the Construction Industry in the 21st Century* (*Niju'uisseiki eno Kensetsu Sangyo Bijyon*), Tokyo: Ministry of Construction, and Construction Industry Vision Study Group, 1986.
2. *Construction Industries of Major Western Countries* (*Shuyo'okoku no Kensetsugyo'o no Genjo'o to Do'oko'o*), Tokyo: Research Institute of Construction and Economy, 1985.
3. *White Paper on Construction* (*Kensetsu Hakusho*), Tokyo: Ministry of Construction, published annually.
4. *Study on International Comparison of Labor Productivity* (*Ro'odo'o Seisansei no Kokusai Hikaku ni Kansuru Kenyu'u*), Tokyo: Japan Productivity Center, 1982.
5. Noritake Kobayashi, *Multinational Corporations of Japan* (*Nihon-no Takokuseki Kigyo'o*), Tokyo: Chuo Keizaisha, 1980.
6. Kazuhiko Kondo, *Construction* (*Kensetsu*), Tokyo: Nihon Keizai Shinbunsha, 1987.
7. Nomura Research Institute, *Growth Strategies of the Construction Industry* (*[Shin] Kensetsu Sangyo no Seicho'o Senryaku*), Tokyo: Seibunsha, 1985.
8. Mitsubishi Research Institute, *Challenge for a New Construction Industry* (*Atarashii Kensetsusangyo'o eno Cho'osen*), Tokyo: Seibunsha, 1986.
9. Pat Choate and Susan Walter, *America in Ruins*, Translated by Social Capital Study Group, Washington, D. C. : The Council of State Planning Agency, 1981.
10. *Monthly of Construction Statistics* (*Kensetsu To'okei Geppo'o*), Tokyo: Ministry of Construction, and Research Council for Prices in the Construction Industry, January 1985.
11. *Monthly Report on Construction Labor and Materials* (*Kensetsu Ro'odo'o Shizai Geppo'o*), Ministry of Construction, Tokyo: Taisei Shuppansha, January 1985.
12. *The Construction Industry in the 21st Century* (*21seiki no Kensetsu Sangyo*), Tokyo: Shimuzu Corporation, 1980.
13. *Nikkei Architecture*, Tokyo: Nihon Keizai Shinbunsha, July 13, 1987.
14. *Construction Directory 1988* (*Kensetsu Meikan 1988*), Tokyo: Nikkan Kensetsu Tsushinsha, 1988.
15. *Economic Statistics Annals* (*Keizai Tokei Nenkan*), Tokyo: Toyo Keizai Shinposha, 1968-1983.
16. *Analysis of Labor Demand Factors in the Construction Industry* (*Kensetsugyo'o ni okeru Ro'odo Juyo'o Yo'oin no Bunseki*), Tokyo: Mitsubishi Research Institute Inc., 1980.
17. "Survey Report on Demand and Supply of Skilled Labor" (*Gino'o Ro'odo'osha Jukyu'u Jyo'okyo'o Cho'osa Ho'okokusho*), *Statistical Annals,* Tokyo: Ministry of Labor.

18. *Labor Force Survey* (*Ro'odo'oryoku Cho'osa*), Tokyo: Management and Coordination Agency.

19. Kazuo Shimotaka, "Long-Term Guarantee on Quality by Japanese Housing and Urban Development Corporation" (*Jyuto Ko'odan ni okeru Seino'o Hattyu Ho'oshiki to Hinshitsu no Cho'oki Hosyo'o*), in *Construction Work* (*Seko'o*), Tokyo: Syo'okokusha, July 1985.

20. *Survey Report on Orders Received from Overseas Construction Work* (*Kaigai Kensetsukoji Juchu'u Cho'osa*), Tokyo: Ministry of Construction, 1986.

21. "Top International Contractors," *Engineering News Records*, McGraw-Hill, July 17, 1986.

22. Nomura Research Institute, *Basic Directions and Corresponding Measures for Overseas Construction* (*Kaigai Kensetsu no Kihon Ho'oko'o to Taio'osaku*), Tokyo: Overseas Construction Association of Japan Inc., 1984.

23. *Monthly Ohbayashi*, Tokyo: Ohbayashi Corporation, March 1985.

24. *Nikkan Kensetsu Sangyo Shinbun*, Tokyo: June 23, 1986.

25. Corporate Annual Reports of Shimizu Corporation for fiscal 1973, 1983 and 1986.

26. Corporate Annual Reports of Kajima Corporation, Takenaka Corporation and Kumagai Corporation for fiscal 1984, 1985 and 1986.

INDEX